Modern Gastronomy

A to Z

a Scientific and Gastronomic Lexicon

Modern Gastronomy A to Z

a Scientific and Gastronomic Lexicon

Alícia Foundation
elBullitaller

Alícia Foundation founded by
famed chef Ferran Adrià of el Bulli

CRC Press
Taylor & Francis Group
Boca Raton London New York

CRC Press is an imprint of the
Taylor & Francis Group, an **informa** business

THE CULINARY INSTITUTE OF AMERICA®
THE WORLD'S PREMIER
CULINARY COLLEGE

CRC Press
Taylor & Francis Group
6000 Broken Sound Parkway NW, Suite 300
Boca Raton, FL 33487-2742

Printed in the United States of America on acid-free paper
10 9 8 7

International Standard Book Number: 978-1-4398-1245-7 (Hardback)

Visit the Taylor & Francis Web site at
http://www.taylorandfrancis.com

and the CRC Press Web site at
http://www.crcpress.com

Contents

Foreword

This book, the *Modern Gastronomy: A to Z*, is a landmark in the history of cooking.

It is not a grand monument. It's a concise, factual handbook. And it's not definitive. It will be superseded by improved editions of itself. But this modest book does mark the arrival of a new era for the culinary profession. In recent years, a few leading chefs have begun to take a fresh look at their craft and explore new ways of bringing nourishment, pleasure, and meaning to our lives. *Modern Gastronomy: A to Z* defines the chemical materials and processes that are the basis for the craft of cooking. It's not the first book to present this kind of information. But it *is* the first such book to be initiated and shaped by professional cooks themselves, and to be so forward-looking. Many of the entries describe materials and methods whose current use is "in experimentation." The book of cooking is now wide open.

Though the combination of cooking and science may seem fashionably modern, in fact these two disciplines go way back together. Cooks were among the first practical chemists on the planet. They discovered through trial and error that we could use tools, heat, and fermentation to transform natural foodstuffs into safer, more nutritious, and more interesting foods. When true experimental science developed in the 17th century, the early chemists learned important lessons from cooks and their water extracts of plant and animal tissues—their soups and stocks. As the knowledge of food chemistry grew, a number of scientists came to write about cooking and food preparation, among them Benjamin Thompson, Justus Liebig, and Louis Pasteur.

In the 20th century, the needs of the expanding industrial sector gave rise to the specialized field of food science. I began to survey the food science literature in 1978, and distilled the information relevant to cooks in a 1984 book, *On Food & Cooking: The Science & Lore of the Kitchen*. Then, in 1992, came the first of a handful of biannual workshops on "molecular gastronomy" in Erice, Sicily. The idea of an Erice workshop on the science of cooking was proposed by California cooking teacher Elizabeth Cawdry Thomas, and organized by her friend University of Oxford physicist Nicholas Kurti, who then invited Hervé This and myself to assist him. Though the term "molecular gastronomy," coined by Kurti and This, suggests precise, state-of-the-art analysis (as in its model, molecular biology), in fact the Erice meetings brought together a very mixed group of scientists and chefs, and focused on the basic food chemistry of traditional dishes. The presentations and discussions were informal, and were never published.

Now, at the beginning of the 21st century, an experimental, open approach to cooking—often mistakenly named "molecular gastronomy"—has burst into prominence. This approach owes very little to particular books or meetings. It is the product of broad historical developments and a unique catalyst: the globalization of travel and commerce, the expansion of the Internet and rapid access to information, the infiltration of modern technology into all aspects of life—and the vision of Ferran Adrià.

In the past, cooks and their creations have been constrained by many factors—the limited local and seasonal availability of ingredients, limited techniques and tools for transforming them, limited understanding of cooking processes, and the necessarily narrow expectations defined by local customs. Of course, limitation can inspire inventiveness, and this is what has given us our great culinary traditions. But today there are many fewer constraints, and therefore unprecedented opportunities for the craft of cooking to grow and advance. Imaginative cooks are now able to work with the entire planet's ingredients, cooking methods, and traditions, and draw on all of human knowledge and invention to explore what can be done with food and the experience of eating.

Several remarkable chefs have been at the forefront of this new open cuisine, but its chief trailblazer is Ferran Adrià. In 1987, soon after rising to head the kitchen at El Bulli, a small restaurant hours north of Barcelona, he began a deliberate process of culinary exploration that has been dazzlingly productive and influential. He and his team became prolific creators of startling, beautiful, delicious dishes that brought worldwide renown to El Bulli and to elBullitaller, the Barcelona workshop that he set up in 1997. Science and scientists came relatively late to the workshop, but are now an integral part of its creative work. In 2004, Ferran Adrià agreed to direct a joint undertaking of the Manresa Savings Bank and the Catalan regional government, the new Alicia Foundation, whose name reflects the meeting of food (Catalan *alimentació*) and science (*ciència*). Alicia's purpose is to foster collaboration among culinary professionals, food scholars, scientists, and educators, to improve the quality of food—and so the quality of life—for as many people as possible.

This book is the first major project of the Alicia Foundation, a joint publication with elBullitaller. It is both a gift and a challenge. The gift is the book itself and the information it holds. The challenge is to make use of it creatively and generously, in the same spirit that has brought it to your hands.

Harold McGee

Preface

Despite the fact that most reactions and techniques in the kitchen have a scientific explanation, science and gastronomy have joined forces on very few occasions. Recently, there have been initiatives to establish a debate between both disciplines which, in principle, have very little in common with each other insofar as objectives and methods are concerned.

The reason for this is simple; these initiatives have arisen from an awareness that has gradually been taking shape in the world of gastronomy: knowledge of the processes that make culinary operations possible cannot but benefit all professionals who have hitherto used the traditional method of trial and error.

Modern Gastronomy: A to Z aims not only to give chefs a better understanding of the terminology that describes the nature of the ingredients they work with every day and to explain the reasons these ingredients produce certain reactions, but also to help them discover the potential of a wide range of products that can be used in a diversity of preparations.

This is a book that aims to enable catering professionals to research quickly, easily, and in plain language everything they need to know with regard to science in cooking. The choice of entries was above all practical; we hope, therefore, that chefs will find answers to some of their questions and broaden their understanding of a subject that they broach each day with their own hands.

However, when writing this book, the only tools at our disposal have been our thoughts, intuition, and continuous research in widely diverse areas, as there have been very few books to consult on this topic. Therefore, more than ever, we feel bound to state that this is a living work that sees itself as the foundation stone, so that in the future, all of us, the scientists who have been involved in the work, as well as the chefs who use it, will improve and adapt it to the real needs of the world of gastronomy.

Pere Castells
Director of the Scientific Department
Fundación Alícia

Albert and Ferran Adrià
elBullitaller

This Lexicon is a living work that aims to adapt and readapt to the real needs of chefs. Please help us to complete and improve it by sending your suggestions.

lexic@alimentacioiciencia.org

alícia

Acknowledgments

The contribution made by both science and gastronomy professionals has determined the shape of *Modern Gastronomy: A to Z*. First, we would like to thank the Roca brothers and Gabriel and Maria; also Salvador Brugués, Andoni Luís Aduriz, Quique Dacosta and Carme Ruscalleda, who agreed to proofread the first drafts. Their comments were of enormous value. The revision of the English version was by Wylie Dufresne.

We received invaluable help from Fernando Sapiña, Joaquín Pérez Conesa, and Raimundo García del Moral from the scientific world. Likewise, Robert Xalabarder and Claudi Mans edited the book and helped to correct the contents. The scientific adaptation in English was done by Ramon Trujillo (University of London) and Héloise Vilaseca. Josep Maria Pinto structured and gave shape to all the contents. And finally, we would like to thank Ingrid Farré, Elena Roura, and Toni Massanés at the Fundación Alícia, as well as all the team at elBulli for making possible this idea that we nurtured for so long.

Pere Castells
Director of the Scientific Department
Fundación Alícia

Albert and Ferran Adrià
elBullitaller

Scientific review and technical editing of the translated manuscript by

Darryl L. Holliday
Department of Food Science
Louisiana State University
Baton Rouge, Louisiana

How to Use *Modern Gastronomy*

How does *Modern Gastronomy* work?

1. As its name would suggest, the book has a lexical format with all the entries in alphabetical order.
2. A definition is given in each entry, which helps to establish each word quickly.
3. The "additional information" that accompanies the entries allows the reader to study the contents in greater depth and to understand the entry.
4. In the case of additives and other products, we offer practical suggestions for use in the food and catering industries, as well as instructions for dosage and use.
5. When it is indicated that a product is found "in experimentation," this does not mean that it is not yet used in cooking. In fact, many of these products have been used in restaurants, but we prefer not to deal with them in depth in this book until we have more information. We hope that they will be included in later editions.
6. So that each entry is situated in the right field, we have divided them into different categories (see table below). The category or family is included in small gold print just beneath the entry heading.
7. Naturally, some entries belong to more than one category. For example, acetic acid is included in the food composition—acids family, as well as in the additives—acidity regulators and additives—preservatives families.

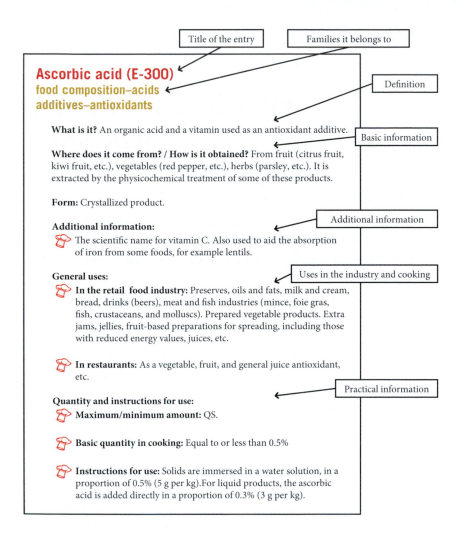

Title of the entry

Families it belongs to

Ascorbic acid (E-300)

food composition–acids
additives–antioxidants

Definition

What is it? An organic acid and a vitamin used as an antioxidant additive.

Basic information

Where does it come from? / How is it obtained? From fruit (citrus fruit, kiwi fruit, etc.), vegetables (red pepper, etc.), herbs (parsley, etc.). It is extracted by the physicochemical treatment of some of these products.

Form: Crystallized product.

Additional information

Additional information:

- The scientific name for vitamin C. Also used to aid the absorption of iron from some foods, for example lentils.

Uses in the industry and cooking

General uses:

- **In the retail food industry:** Preserves, oils and fats, milk and cream, bread, drinks (beers), meat and fish industries (mince, foie gras, fish, crustaceans, and molluscs). Prepared vegetable products. Extra jams, jellies, fruit-based preparations for spreading, including those with reduced energy values, juices, etc.

- **In restaurants:** As a vegetable, fruit, and general juice antioxidant, etc.

Practical information

Quantity and instructions for use:

- **Maximum/minimum amount:** QS.

- **Basic quantity in cooking:** Equal to or less than 0.5%

- **Instructions for use:** Solids are immersed in a water solution, in a proportion of 0.5% (5 g per kg).For liquid products, the ascorbic acid is added directly in a proportion of 0.3% (3 g per kg).

Additives	
Antioxidants	Stabilizers
Coloring agents	Gases
Preservatives	Gelling agents
Sweeteners	Humectants
Emulsifiers	Flavor enhancers
Thickening agents	Acidity regulators
Food composition	
Acids	Lipids
Alkaloids	Minerals
Alcohols	Pigments and other compounds
Carbohydrates	Proteins
Food concepts	
Scientific concepts	
Organoleptic perceptions	
Physical or chemical processes	
Mineral products	
Technology	
Devices	Utensils

A

Acacia gum

See **Arabic gum**

Acesulfame-K (E-950)

additives—sweeteners

What is it? A potassium salt used as a sweetening additive, derived from an amide (an organic compound characterized by the presence of nitrogen).

Where does it come from? How is it obtained? Obtained by synthesis in the chemistry industry from petroleum derivatives (acetoacetamide).

Form: Powder.

Additional information:

- It is an intensive sweetener, also called potassium acesulfame or ace-K.
- Its sweetening power is some 200 times greater than sucrose (sugar).
- If stored in appropriate cold and dry conditions, it has a useful life of approximately 4 years.

General uses:

- **In the retail food industry:** Confectionery, drinks, chewing gum, jams. Sweetener for diabetics.
- **In restaurants:** Sugar-free drinks and desserts.

Acetic acid (E-260)

food composition—acids
additives—acidity regulators
additives—preservatives

What is it? An organic acid, a component of vinegar and an additive when it is used in its pure state. It is used as an acidity regulator and a preservative.

Where does it come from? How is it obtained? Produced by the fermentation of diluted alcoholic liquids, typically wine, cider or beer and other products (sake—rice wine, etc.).

Form: Liquid.

Additional information:

- It is found in a proportion of approximately 6% in vinegar. Pickling generally employs more concentrated solutions.

General uses:

- **In the retail food industry:** Vinaigrettes, sauces, cheeses, special breads, etc.
- **In restaurants:** No direct application. Used as vinegar when dissolved.

Acid

scientific concepts
organoleptic perceptions

What is it?

- **Scientifically:** A product with an organoleptic characteristic caused by its tendency to give off hydrogen ions.
- **Gastronomically:** The name given to one of the basic tastes.

Additional information:

- Acidic products have an acidity that can range from pH 0 to pH 7 (in food, from 2.5 to 7). *See* pH.

General uses:

- In cooking, the acidic taste can be accentuated by citrus fruits (citric acid), fermented products (acetic acid) or by malic acid from apples (tart tasting).

Most Frequently Used Acids in the Food Industry		
Fruit Derivatives	**Fermentation Derivatives**	**Others**
Citric acid (oranges, lemons, etc.)	Acetic acid (vinegar)	Ascorbic acid (vitamin C)
Malic acid (apples, etc.)	Butyric acid (cheeses)	Phosphoric acid (soft drinks)
Tartaric acid (grapes, etc.)	Lactic acid (yogurt)	

Acidification
physical or chemical processes

What is it? A procedure in which an acid is added to a product or a preparation.

Acidifier
food concepts

What is it? A product that increases the acidity of food and gives it an acid taste.

Additional information:

- Favorable conditions are created for some antioxidants when acidity levels are increased in a product. This has several repercussions, one of which is the delay in the loss of color in fruits and vegetables: Citric, malic, acetic, lactic acid, etc. Products with these components: lemon juice, vinegar, etc.

General uses:

- **In the retail food industry:** Citric acid, acetic acid, tartaric acid, etc. They are preservatives and enhance the taste of vegetables, sauces, jams, sorbets, ice creams, soft drinks, marinades, and acid preparations in general.
- **In restaurants:** Vinegar, lemon juice, etc. For making vinaigrettes, sauces, etc.

Acidity
scientific concepts

What is it? The indication of the acidic strength of a product.

How is it measured? Scientifically measured in pH units. The margin is from 0 to 14 for watery solutions. However, the acidity is relevant only for pH under 7. The range between 7 and 14 indicates the basicity or alkalinity of a product.

Additional information:

- Stomach acidity or heartburn is the name given to a sensation of pain in the abdomen. It is caused by excess gastric production of hydrochloric acid. It can be treated with products that neutralize this acid, such as bicarbonate.
- The following FDA definitions apply:
  - (a) *Acid foods* means foods that have a natural pH of 4.6 or below.
  - (b) *Acidified foods* means low-acid foods to which acid(s) or acid food(s) are added; these foods include but are not limited to beans, cucumbers, cabbage, artichokes, cauliflower, puddings, peppers, tropical fruits, and fish, singly or in any combination. They have a water activity (aw) greater than 0.85 and have a finished equilibrium pH of 4.6 or below. These foods may be called, or may purport to be, "pickles" or "pickled." Carbonated beverages, jams, jellies, preserves, acid foods (including such foods as standardized and nonstandardized food dressings and condiment sauces) that contain small amounts of low-acid food(s) and have a resultant finished equilibrium pH that does not significantly differ from that of the predominant acid or acid food, and foods that are stored, distributed, and retailed under refrigeration are excluded from the coverage of this part.
  - (c) *Low-acid foods* means any foods, other than alcoholic beverages, with a finished equilibrium pH greater than 4.6 and a water activity (aw) greater than 0.85. Tomatoes and tomato products having a finished equilibrium pH less than 4.7 are not classed as low-acid foods.

General uses:

- **In the retail food industry:** Acidity is used to determine the extent of processing a food product needs, i.e., low acid foods need to be processed at a higher temperature for a longer period of time to ensure safety because the acid acts to inhibit some microbial growth.
- **In restaurants:** Acidity has a greater role in flavor than cooking.

Indication of the Acidic Strength of a Product		
Degree of Acidity	**pH**	**Examples**
Very acidic product	Less than 3.5	Lemon juice (2.5)
Acidic product	Between 3.5 and 5	Tomato juice (4.5)
Less acidic product	Between 5 and 7	Melon (6.5)
Neutral product	Equal to 7	Water
Base product	Between 8 and14	Egg white (8.9)

Additive (food)

food concepts

What is it? In the United States, *food additives* include all substances not exempted by section 201(s) of the Code of Federal Regulations, the intended use of which results or may reasonably be expected to result, directly or indirectly, either in their becoming a component of food or otherwise affecting the characteristics of food. A material used in the production of containers and packages is subject to the definition if it may reasonably be expected to become a component, or to affect the characteristics, directly or indirectly, of food packed in the container. "Affecting the characteristics of food" does not include such physical effects, as protecting contents of packages, preserving shape, and preventing moisture loss. If there is no migration of a packaging component from the package to the food, it does not become a component of the food and thus is not a food additive. A substance that does not become a component of food, but that is used, for example, in preparing an ingredient of the food to give a different flavor, texture, or other characteristic in the food, may be a food additive.

In the European Union, food additive means any substance not normally consumed as a food by itself and not normally used as a typical ingredient of the food, whether or not it has nutritive value, the intentional addition of which to food for a technological (including organoleptic) purpose in the manufacture, processing, preparation, treatment, packing, packaging, transport or holding of such food results, or may be reasonably expected to result (directly or indirectly), in it or its byproducts becoming a component of or otherwise affecting the characteristics of such foods. The term does not include contaminants or substances added to food for maintaining or improving nutritional qualities.

A substance with no nutritional value on its own that is deliberately added to a product or food preparation to ensure its preservation, to simplify or improve its preparation process, or to modify its physical or organoleptic properties.

Additional information:

- A nomenclature with an E followed by three or four numbers is assigned to additives in the European Union.
- Additives may be natural or artificial. The E nomenclature does not define this particular aspect.
- They are classified as intermediate food products.
- They have been used quite a bit in the retail food industry but not a lot in kitchens due to their rather bad reputation, except MSG.
- Every country across the globe has a list of authorized additives. In the European Union 365 have been catalogued and the industry regularly uses approximately 125.
- They are "licensed" with the letter E (for Europe) and a three- or four-digit number. When the number E-1420 is marked on a label, this does not mean there are more than 365 additives. Rather, there are many blank spaces on the list to separate the various functions.
- The following functions are the most common: preservative, antioxidant, thickening agent, emulsifier, coloring agent, etc.
- The use of additives, besides contributing to the evolution of the world of textures, is one of the key points in the dialogue between science, food, and gastronomy.

Aerosol

scientific concepts

What is it? A mixture composed of a solid or liquid included in a gas.

Additional information:

For example, the smoke produced when food is cooked over a wood-burning grill contains solid and liquid particles that fly up into the air and come into contact with the food. This gives certain characteristics to this particular cooking process.

Types of aerosol:

- **Liquid aerosol** (L/G) "Water in gas." L indicates liquid and G represents gas. Example: fog, where air disperses little drops of water.
- **Solid aerosol** (S/G) "Solid in gas." The S indicates solid and G represents gas. Example: smoke, where air disperses carbon particles or ash. Smoked foodstuffs are an example of a cooking application of solid aerosols.

The name aerosol has been used to denote the device that generates this kind of dispersion, much used in perfumes, pesticides, etc. In food it is used to disperse aromas and also as a propellant for food additives and food, i.e., pressurized nitrogen or carbon dioxide for whipped cream, or nonstick spray.

Agar-agar (E-406)

additives–gelling agents

What is it? A fibrous carbohydrate that is used as a gelling agent. It has the properties of a hydrocolloid.

Where does it come from? How is it obtained? It is extracted by physico-chemical treatments from red algae *Gelidium* and *Gracilaria.*

Form: Product in powder or in filaments (dehydrated algae).

Additional information:

- It forms thermoreversible gels. *See* Thermoreversibility.

General uses:

- **In the retail food industry:** Baking, preserved vegetable products (preserves, jellies, jams, etc.), meat derivatives, ice cream, cottage cheese, coatings for fish preserves and semi-preserves, soups, sauces, marzipans, fruit-based preparations for spreading, etc.
- **In restaurants:** Jellies. Used for the first time to obtain hot jellies in 1998. Emerging uses include desserts and "noodles."
- **Other:** It is used in science as a solid medium to cultivate microorganisms.

Quantity and instructions for use:

- **Maximum/minimum quantity:** QS (minimum needed to obtain the desired effect), except in jams or industrial derivatives where a maximum of 10g/kg is specified (separately or in total). This means that the agar may be mixed with other hydrocolloids but the total may not surpass 10 g/kg.
- **Basic quantity for cooking:** 0.2–1.5%, i.e., 0.2–1.5 g per 100 g of liquid to be jellified (2–15 g per kg).
- **Instructions for use:** It is stirred and heated to 80°C (for ease it may be heated to boiling point). Jellification begins at between 50 and 60°C and once jelled it can be served hot. The gel withstands temperatures up to 80°C. If a more consistent gel is required, the dosage should be increased, and diminished if a more fluid gel is required.

Algae

Seaweed has been a part of human nutrition for many centuries, both for consumption and for its derivative products. Of the most commonly used algae, a distinction needs to be made between those used in restaurants (as well as in domestic cooking) and those used by the retail food industry to obtain gelling and thickening agents and stabilizers. Some of these algae, such as agar-agar or the carrageenans, are mainly used in the processing industry to obtain additive products (agar-agar or the kappa, iota, and lambda carrageenans), but in some parts of the world they are used for direct consumption.

Common Algae for Direct Consumption
Brown or Phaeophyta algae • Nori (*Porphyra*) • Wakame (*Undaria*) • Kombu (*Laminaria*) • Others: Hiziki, arame, alaria, etc.
Red or Rhodophyta algae • Dulse (alga *Palmaria*) • Agar-agar (alga *Gelidium*) • Carrageen, incorrectly called Irish moss (*Chondrus*)
Blue or Cyanophyta algae (micro-algae) • Spirulina (*Spirulina*) • Sea spaghetti (*Himanthalia*)
Green or Chlorophyta algae • *Caulerpa* • Sea lettuce (*Ulva*)

Common Algae Used to Obtain Additive Products
Brown or Phaeophyta algae • Extraction of alginates: Algae that contain algine or alginate (*Ascophyllum nodosum, Laminaria digita, Fucus serratus, Macrocystis pyrifera,* etc.)
Red or Rhodophyta algae • Extraction of agar-agar. *Gelidium*, in particular *Gelidium sesquipedale* • Extraction of the kappa, iota, and lambda carrageenans. *Chondrus crispus, Gigartina radula, Euchema cottoni,* etc. • Extraction of the furcellaran-type carrageenans. *Furcellaria fastigiata*

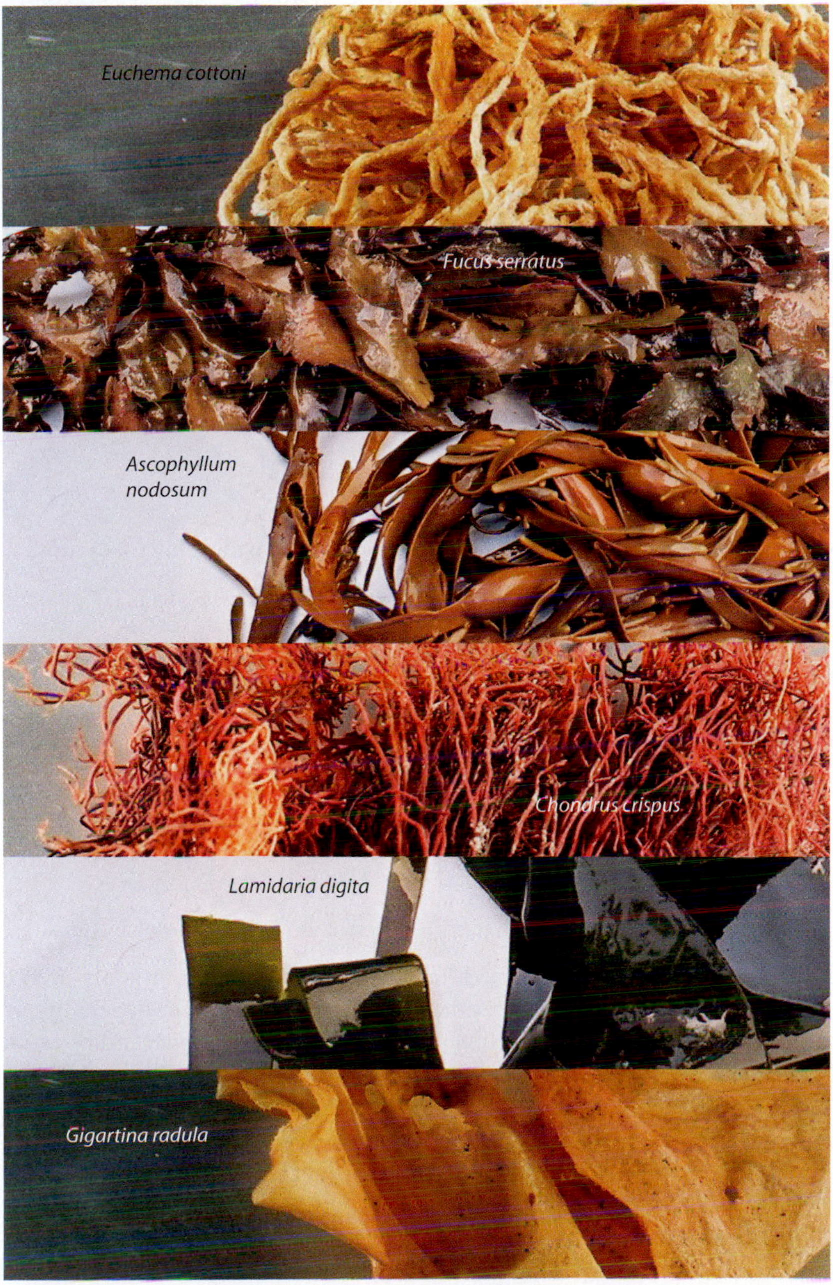
Euchema cottoni
Fucus serratus
Ascophyllum nodosum
Chondrus crispus
Lamidaria digita
Gigartina radula

Albumins

food composition—proteins

What are they? Complex, globular-type proteins that form part of certain foodstuffs: egg white (ovalbumin), milk (lactoalbumin), blood (seroalbumin), etc. They can be used as gelling agents and emulsifiers.

Where do they come from? How are they obtained? From the separation of food components (eggs, milk, blood, etc.).

Form: Powder (ovalbumin, lactoalbumin, seroalbumin) or frozen (ovalbumin).

Additional information:

They are soluble in water and coagulate with heat, acids, enzymes, etc. They can also be found in muscles and other animal substances as well as in many vegetable tissues. They are capable of carrying particles in suspension, which is why they are used to clarify.

General uses:

- **In the retail food industry:** Clarification of wines, juices, syrups, etc.
- **In restaurants:** The egg white albumin is used to clarify stocks. Its emulsifying and gelling properties are used in cooking (mayonnaise, meringues, custards, etc.). Curdling occurs between 70 and 90°C or through acidification or enzymes.

Alcoholmeter

technology—devices

What is it? A device that determines the alcoholic strength of a liquid.

How does it work? There are different types. The most commonly used is an adaptation of a densimeter: it is placed in a liquid and, depending on the quantity of alcohol, is submerged to a depth that directly indicates the alcoholic strength (in % of volume).

General uses:

- **In the retail food industry:** In the wine and distillery industries, flavor industry, biodiesel, etc.
- **In restaurants:** In experimentation or in-house distillation.

Alcohols

scientific concepts
food composition—alcohols

What are they?

- **Scientifically:** Products composed of carbon chains where the main characteristic is the hydroxyl group (part of an organic molecule formed by oxygen and hydrogen).
- **Gastronomically:** A name associated to some types of the clear, colorless fruit brandy known as eau-de-vie, as well as fermented grain spirits.

Additional information:

- The following elements from the world of food belong to this group:
  - Ethyl alcohol/ethanol commonly referred to as "alcohol."
  - The polyol group (caloric sweeteners).
  - Glycerine, which is combined with fatty acids to produce glycerides, which in turn make up lipids.
- Of non-foodstuffs, methyl alcohol/methanol is cheaper than ethanol and has sometimes been used fraudulently to prepare alcoholic beverages. This type of usage has given rise to very dangerous cases of intoxication.

Alginates

additives—gelling agents
additives—thickening agents
additives—stabilizers

What are they? Organic salts derived from fibrous carbohydrates used as gelling and thickening agents and stabilizers. They have the properties of hydrocolloids.

Where do they come from? How are they obtained? They are extracted by physicochemical treatments of brown algae (*Macrocystis, Fucus, Laminaria ascophyllum,* etc.) that are found in cold-water seas and oceans.

Additional information:

The name is derived from "alga." Of the approximately 30,000 species that make up this group of generally aquatic plants, only 50 are used for human nutrition.

Most common alginates: Sodium, potassium, calcium, ammonium, and propylenglycol salts of alginic acid. The retail food industry mainly uses sodium alginate (E-401) and the propylenglycol alginate (E-405), the latter as a foam regulator, for example, in certain types of beer. *See* Sodium alginate.

Alimentary
food concepts

What is it? Everything relating to nourishment or nutrition; furnishing sustenance or maintenance.

Alkali
scientific concepts

What is it? A product with chemical properties that are opposite to acidic properties. It has a definite organoleptic character (soapy) caused by its tendency to capture hydrogen ions. Its pH is above 7 up to 14. It is also referred to as a base.

Additional information:

It is used to diminish the acidity of some foods (acid neutralizers). Examples: bicarbonate of soda, sodium citrate, calcium carbonate, etc. There are very few foodstuffs that are slightly alkaline, such as eggs, soda crackers, and cocoa.

Examples: Sodium hydroxide (caustic soda, E-524), ammonium (E-527), bicarbonate of soda (E-500), sodium carbonate (soda ash, E-500i), sodium citrate (E-331), etc.

General uses:

- **In the retail food industry:** Treatment of olives with caustic soda (0.25–2%) to eliminate the bitter taste and to produce a dark color. Immersion of breads and cakes in caustic soda (1.25%) at 85–90°C to produce a brownish color on the surface. Bicarbonate of soda is used in the manufacture of chocolate to enhance the Maillard reaction and thereby increase the bitter taste and dark color, etc. Also used in cleaning and sanitizing compounds.
- **In restaurants:** Sodium citrate is used to reduce acidity in acidic products and to enable gelling or thickening actions in processes such as spherification. Also used in cleaning and sanitizing compounds. *See* Sodium alginate.

Alkaloids
food composition—alkaloids

What are they? Organic products contained in some vegetables, often with stimulating properties.

Examples: Nicotine in tobacco, morphine in opium, cocaine in the coca plant. In the retail food industry: chaconine and solanine in potatoes, caffeine in coffee and tea.

Additional information:

- They normally have a bitter taste and once extracted from the vegetable, they are solid and colorless.
- Alkaloids have different effects on the organism that can often be harmful. Only caffeine is directly applied in the retail food industry. *See* Caffeine.

Aluminum (E-173)
additives—coloring agents

What is it? A metallic chemical element with a silvery color, used as a surface coloring agent.

Where does it come from? How is it obtained? In chemical processes using aluminum salts.

Form: Powder.

Additional information:

Besides being used as a coloring agent, it is widely used in the form of aluminum foil.

Note: "Silvery" might mean that it has an appearance like that of silver, and not necessarily that it is made of silver.

General uses:

- **In the retail food industry:** A coloring agent for the surfaces of confectionery products. It can be used in the manufacture of cookware. It is used in the canning industry to preserve color or texture, also used as an anti-caking agent.
- **In restaurants:** Aluminum foil. Found in many products from processed cheese to baking soda.

Amino acids
food composition—acids

What are they? All simple biochemical products with both amine and carboxylic acid groups.

Where do they come from? How are they obtained? By physicochemical treatment of protein decomposition; however, it is more usual to obtain them from biofermentation.

Form: Powder.

Additional information:

- Products with simple structures. The presence of nitrogen distinguishes them from other components of biological material (carbohydrates, lipids, etc.).
- There are 20 amino acids in the natural world, nine of which are essential; in other words, the only way they can come into contact with the organism is through food (valine, leucine, isoleucine, histidine, threonine, methionine, lysine, phenylalanine and tryptophan). The others are produced by the organism.
- The 20 basic amino acids that compose proteins are as follows:

Name	Symbol
Alanine	Ala (A)
Arginine	Arg (R)
Asparagine	Asn (N)
Aspartic acid	Asp (D)
Cysteine	Cys (C)
Glutamine	Gln (Q)
Glutamic acid	Glu (E)
Glycine	Gly (G)
Histidine	His (H)
Isoleucine	Ile (I)
Leucine	Leu (L)
Lysine	Lys (K)
Methionine	Met (M)
Phenylalanine	Phe (F)
Proline	Pro (P)
Serine	Ser (S)
Threonine	Thr (T)
Tryptophan	Trp (W)
Tyrosine	Tyr (Y)
Valine	Val (V)

Some sources include hydroxylisin and hydroxyproline, which brings the list of amino acids to 22.

Examples of amino acid protein percentages in some foodstuffs (g per 100 g of total proteins). Obviously, the total percentage of proteins in one foodstuff is used to make comparisons.

	Veal	Eggs	Cow's Milk	Peas	Cod
Proteins	20.3%	12.5%	3.2%	5.8%	17.4%
Ile	4.9%	5.6%	4.9%	4.7%	5.2%
Leu	7.6%	8.3%	9.1%	7.5%	8.3%
Lys	8.7%	6.3%	7.4%	8.0%	9.6%
Met	2.6%	3.2%	2.6%	1.0%	2.8%
Cys	1.2%	1.8%	0.8%	1.2%	1.1%
Phe	4.3%	5.1%	4.9%	5.0%	4.0%
Tyr	3.7%	4.0%	4.1%	3.0%	3.4%
Thr	4.5%	5.1%	4.4%	4.3%	4.7%
Trp	1.2%	1.8%	1.3%	1.0%	1.1%
Val	5.1%	7.6%	6.6%	5.0%	5.6%
Arg	6.4%	6.1%	3.6%	1.0%	6.2%
His	3.5%	2.4%	2.7%	2.4%	2.8%
Ala	6.1%	5.4%	3.6%	4.5%	6.7%
Asp+Asn	9.1%	10.7%	7.7%	11.9%	10.2%
Glu+Gln	16.5%	12.0%	20.6%	17.3%	14.8%
Gly	5.6%	3.0%	2.0%	4.3%	4.6%
Pro	4.9%	3.8%	8.5%	4.1%	4.0%
Ser	4.3%	7.9%	5.2%	4.7%	4.8%

Adapted from Coultate, T.P. (2002). *Food: The Chemistry of Its Components*, Fourth Edition. The Royal Society of Chemistry, Cambridge, U.K.

General uses:

- **In the retail food industry:** Food complement, foaming agent, flour conditioners (cysteine), browning agents (leucine), flavor enhancement (glutamine).
- **In restaurants:** MSG (glutamine).

Amylase

food composition—proteins

What is it? An enzyme that creates the breakdown of amylose chains in starch.

Additional information:

- Present in pancreatic juices and saliva, it breaks down the amylose chains of starch (depolymerization). The breakdown causes the formation of sugar-type carbohydrates like glucose. Thus, when flour is held for a while in the mouth, amylase acts in the saliva, resulting in a sweet taste. However, this process is not complete, as glucose and fragments of starch called dextrines remain that may not be broken down.
- The amylopectin is also attacked by the amylase but the bonds cannot be broken down without the intervention of another enzyme: glucosidase.
- These enzymes can be used to obtain glucose chains with sweetening powers that depend on the breakdown process.

General uses:

- **In the retail food industry:** Flour conditioners.
- **In restaurants:** None.

Amylopectin

food composition—carbohydrates

What is it? A component that constitutes approximately 75% of starch on average. It is composed of groups derived from glucose in the form of ramified chains, i.e., it is a highly branched polymer of glucose.

Additional information:

It is one of the sources of carbohydrate that is present in starch. Its branched structure is one of the factors that determine the thickening properties of starch.

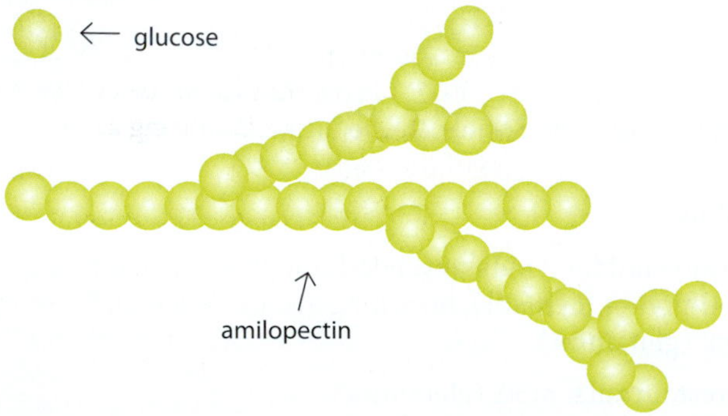
← glucose
amilopectin

Amylose
food composition—carbohydrates

What is it? A component that constitutes approximately 25% of starch on average. It is composed of groups derived from glucose in the form of linear chains, i.e., it is a linear polymer of glucose.

Additional information:
It is one of the sources of carbohydrate that is present in starch. Its linear structure is one of the factors that determine the gelling properties of starch.

glucose
amylose

Anthocyans
food composition—pigments and other compounds

What are they? Products of vegetable origin (flavonoids), responsible for the red (some red comes from lycopene and/or lutein), purple, and blue colors in some foodstuffs.

Where do they come from? How are they obtained? Vegetable extracts.

Form: Usually in liquid form.

Additional information:

- They have a sharp (bitter/astringent) taste.
- The most common foodstuffs that contain anthocyans are red grapes, red cabbage, red onion, berries, etc.
- To preserve their characteristic color, these vegetables should be cooked in slightly acidic liquids, adding a little vinegar, lemon juice, etc. Vegetables with anthocyanins should not be cooked in iron or aluminum containers without a ceramic coating as these materials would alter the colors of the vegetables.
- These pigments are very soluble in water and very sensitive to acidity variations. They are also antioxidants.

General uses:

- **In the retail food industry:** As natural coloring agents.
- **In restaurants:** Coloring agents in the form of fruit juices or vegetable pastes.

Anthoxanthins

food composition—pigments and other compounds

What are they? White, cream, and orangey pigments responsible for the color of vegetables such as cauliflower, onion, etc.

Additional information:

- When in contact with iron and aluminum, they turn foodstuffs a yellowy-brown color. This needs to be borne in mind when using pots without a ceramic coating.
- They have a slightly sharp taste, similar to anthocyans.

Anti-caking agent

food concepts

What is it? A product that reduces the tendency of products (powder, lumpy products, and even solid masses) to stick together over time and in a humid atmosphere.

Additional information:

- One of the most widely used in the retail food industry is silicon oxide (E-551).
- In the past, grains of rice were placed with salt in saltcellars in order to stop the salt from sticking together; nowadays, salt is treated with an anti-caking agent.

General uses:

- **In the retail food industry:** Products in powder form (flour, starch, sugars, salt, etc.).
- **In restaurants:** Corn starch is commonly used to prevent pastries, gummies, or seasoning blends from sticking together or clumping. Flour is used to prevent raw dough from sticking.

Anti-foaming agent

food concepts

What is it? A product that prevents or reduces the formation of foam.

Additional information:

- For example: dimethylpolysiloxane E-514.
- Some emulsifiers in small doses produce anti-foaming effects, for example, monoglycerides and diglycerides (E-471).

General uses:

- **In the retail food industry:** Juices, beers, etc.
- **In restaurants:** Lipids may be used to prevent starch foams in applications such as large pots of rice or pasta.

Antioxidants
food concepts

What are they? Products that avoid oxidation in vulnerable foodstuffs.

Additional information:

- Antioxidants cannot delay oxidation processes that have already begun; their role is preventive.
- In cooking, oxidation occurs when fats turn rancid, when fruit and vegetables change color, etc. Oxidation may be avoided with these products. Traditionally, resources such as lemon juice or parsley have been used.

Commonly used antioxidants:

- Many antioxidants have been identified in the natural world; some of them (tocopherols or vitamin E) are already used as additives (E-306–309), others such as rosemary, sage, or clove extract are expensive and the flavor of the plant means that their applications are limited.
- The most important antioxidant additives are tocopherols, mentioned above, two artificial products (butylhydroxanysol BHA E-320 and butylhydroxytoluene BHT E-321) and the gallates (E-310–312), which are oak tannins and have the disadvantage of dying a foodstuff a blue color if it contains iron (blue ink used to be iron gallate).
- Vitamin C (ascorbic acid) is also considered an antioxidant and is widely used as such although its action is not the same: rather than inactivating the product, it "kidnaps" the oxygen.

General uses:

- **In the retail food industry:** Canned and bottled preserves, dairy products, fats and oils to stop products from going rancid. Also in the meat industry to avoid oxidation of fats.
- **In restaurants:** Applications of ascorbic acid to avoid the discoloring of fruit and vegetables (lowering of pH may have a greater effect).

Arabic gum (E-414)

additives—thickening agents
additives—stabilizers
additives—emulsifiers

What is it? A fibrous carbohydrate that is used as a thickening agent, emulsifier, and stabilizer. It has the properties of a hydrocolloid.

Where does it come from? How is it obtained? It is from the tree *Acacia senegal* and is exuded (by an incision in the tree trunk) and then subjected to physicochemical treatment.

Form: Powder.

Additional information:

- It has been known about for at least 4,000 years.
- Although its thickening power is much less than other gums it has an emulsifying property that enables essential oils (lemon, orange, etc.) to be incorporated into soft drinks.
- It is also used as a soluble fiber in soups and sauces.
- It is also known as acacia gum and Senegal gum.

General uses:

- **In the retail food industry:** Soups, sauces, wine clarifier, drinks, marzipan, aroma encapsulator, cocoa, and chocolate coating, beer, supplier of soluble fiber, etc.
- **In restaurants:** In experimentation; used in espumas (foams), sauces, and gummies.

Argon (E-938)

additives—gases

What is it? An inert gas from the helium family.

Additional information:

Its use is very limited in the retail food industry (protective atmospheres).

Aroma

organoleptic perceptions

What is it? A perception received by the sense of smell retronasally, when capturing volatile substances providing a particular scent that stimulates the sense of smell.

Artificial products

scientific concepts

What are they? Products that do not exist in the natural world.

Additional information:

All artificial products are obtained by synthesis, and are therefore synthetic. *See* Synthetic products.

Classification:

a. Artificial products obtained from natural products

For example: Neohesperidin DC. A sweetener that is 600 times sweeter than common sugar (sucrose) and is obtained thus:

bitter orange + chemical modification = neohesperidin

b. Artificial products obtained from other artificial products

For example: Saccharin is a sweetener between 400 and 1,000 times sweeter than sucrose and is obtained thus:

synthetic product + chemical reactions = saccharin (toluenesulfonic acid)

Ascorbic acid (E-300)

food composition—acids
additives—antioxidants

What is it? An organic acid and a vitamin used as antioxidant additive.

Where does it come from? How is it obtained? From fruit (citrus fruit, kiwi fruit, etc.), vegetables (red pepper, etc.), herbs (parsley, etc.). It can be extracted by physicochemical treatment of some of these products, but normally we obtain it from chemical synthesis.

Form: Crystallized product.

Additional information:

The scientific name for vitamin C. Also used to aid the absorption of iron from some foods, for example, lentils.

Types of derivatives: Ascorbates (E301, E302) and esters (palmitates and stearates—E-304).

General uses:

- **In the retail food industry:** Canned preserves, oils and fats, milk and cream, bread, drinks (beers), meat and fish industries (mince, foie-gras, fish, crustaceans, and molluscs). Prepared vegetable products. Extra jams, jellies, fruit-based preparations for spreading, including those with reduced energy values, juices, etc.
- **In restaurants:** As a vegetable, fruit, and juice antioxidant, etc.

Quantity and instructions for use:

- **Maximum/minimum quantity:** QS (minimum quantity needed to create the desired effect).
- **Basic quantity for cooking:** Equal to or less than 0.5%
- **Instructions for use:** Solids are immersed in a water solution in a proportion of 0.5% (5 g per kg). For liquid products, the ascorbic acid is added directly in a proportion of 0.3% (3 g per kg).

Aspartame (E-951)

additives—sweeteners

What is it? An artificial product composed of two amino acids (phenylalanine and aspartic acid) and used as an intensive sweetener.

Where does it come from? How is it obtained? It is derived from amino acids. It is obtained by synthesis in the chemical industry.

Form: Powder.

Additional information:

- Its sweetening power is between 180 and 200 times greater than sugar (sucrose).
- It is used in various types of diets.
- To simulate sugar, a proportion of 3.1% aspartame is mixed with maltodextrine, resulting in a product with the same volume and sweetening power as sucrose.
- It contains phenylalanine and is therefore not suitable for phenylketonurics. (Phenylketonuria or ketonuria is a hereditary disease related to intolerance to products that contain the amino acid phenylalanine).

General uses:

- **In the retail food industry:** As a sweetener in drinks, dairy products, cereals, chewing gum, and as sugar substitutes in products for diabetics; widely used as a low-calorie product. Its calorie content is insignificant. It cannot be used in pastry or in processed foodstuffs as it decomposes with prolonged heating.
- **In restaurants:** Sugar-free cold desserts and beverages.

Astringent

organoleptic perceptions

What is it? A taste nuance caused by products that constrict the upper layers of tongue mucus and provoke a sensation of roughness.

Additional information:

- In alcoholic beverages with more than 40%, ethyl alcohol acts as an astringent. The best examples are tannins, which exist in the vegetable world in a wide variety of plants (chestnut, holm oak, pine, grape etc.).
- In the culinary world, the presence of astringents in fruits and derivatives (wine, etc.) defines the flavor. They also give a particular taste to tea, coffee, and cocoa. In fruit, the astringency caused by tannins reduces with ripening.

Atom

scientific concepts

What is it? An elemental component of matter; the smallest unit of a chemical element.

Additional information:

- It is composed of protons, neutrons, and electrons.
- The bonding of atoms creates molecules and ions (parts of a molecule with an electric charge), and the molecules make up the products (matter).

Autoclave

technology—devices

What is it? A device similar to a pressure cooker that allows food in containers to be commercially sterilized. It is also used commercially to sterilize utensils.

How does it work? When the water in the autoclave is heated, the temperature reached by the steam that is created sterilizes the product or utensil deposited in the device.

Additional information:

To guarantee that almost all of the harmful microbes are eliminated, a minimum temperature of 134°C is required at a pressure of 2 bars for 20 to 30 minutes.

General uses:

- **In the retail food industry:** Sterilization of utensils and foodstuffs that are usually packed in glass containers, etc.
- **In restaurants:** No known use.

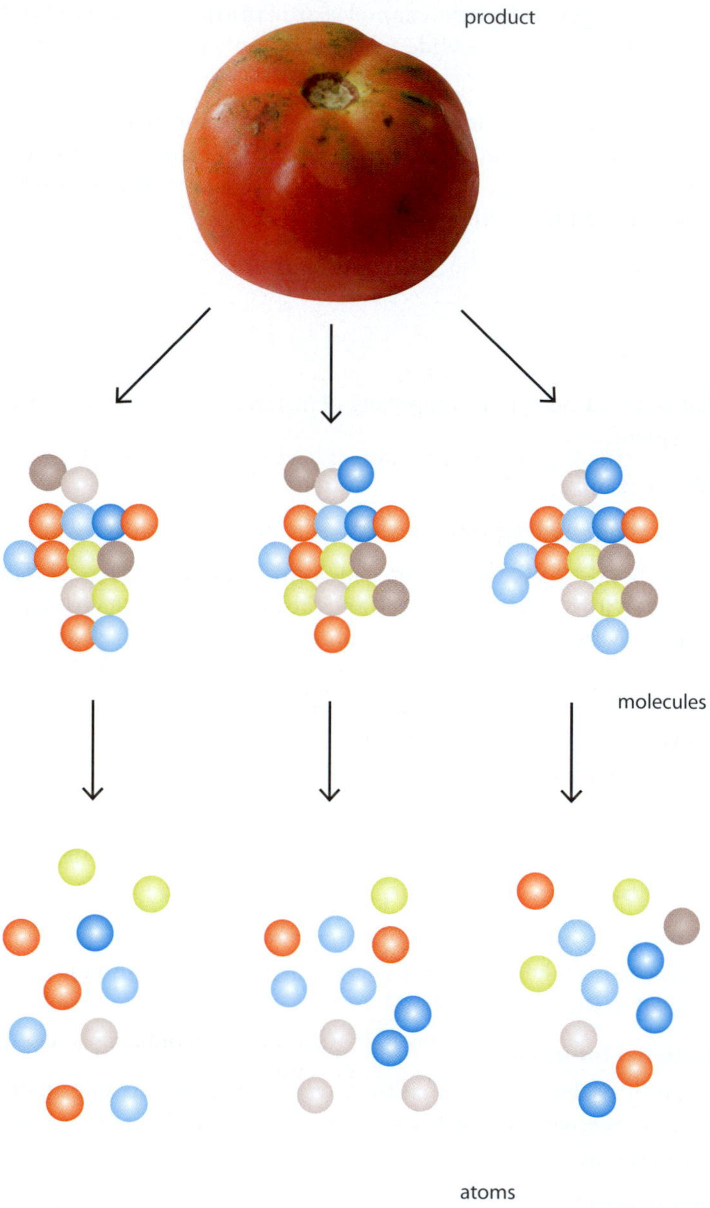
product
molecules
atoms

B

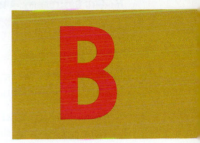

Bacteria
scientific concepts

What are they? Types of microorganisms that have a cell wall, but lack organelles and an organized nucleus.

Where do they come from? / How are they obtained? Laboratory cultivations of bacteria.

Form: Lyophilized powder.

Additional information:

- They exist in a wide variety of mediums and atmospheres.
- Some are harmful, causing diseases (tuberculosis, diphtheria) and are found at the early stages of the most common types of food poisoning (salmonella). Special precautions should be taken with *Clostridium botulinum* as it generates the botulinum toxin that causes the most serious, although rare, cases of food poisoning.
- However, others are beneficial and some traditional foodstuffs are products of their action.

General uses:

- **In the retail food industry:** They are used to initiate controlled fermentations (for example, cheeses, yogurt, etc.).
- **In restaurants:** Care is given to the prevention of harmful bacteria in food. Some use in-house fermentation.

See Microorganism (or Microbe)

Bacteriological incubator

technology—devices

What is it? A device in which microbes may be cultivated in ideal conditions for reproduction and subsistence.

How does it work? In order to stimulate microbiological processes, the temperature and the interior humidity are controlled meticulously in a closed container by a system of sensors.

Additional information:

- Normal devices allow a control at between 5°C and 80°C with a stability of 0.5°C.

General uses:

- **In the retail food industry:** Yogurts and in general fermentation foodstuffs.
- **In restaurants:** No known use.

Base

See **Alkali**

Bentonites

mineral products

What are they? Very absorbent clays used as clarifiers, especially in wine.

Where do they come from? How are they obtained? By treatment of impure clays.

Form: Powder.

Additional information:

- They are added to liquid mixtures.
- They are capable of retaining and then separating the products that produce murkiness in order to achieve a clarified product. This application has led to their being widely used in the wine industry.

General uses:

- **In the retail food industry:** Wine clarification.
- **In restaurants:** No known use.

Bicarbonate of soda (E-500)

mineral products
additives—acidity regulators

What is it? An alkaline product that is used as a food additive. It is the principal component of chemical leaveners.

Where does it come from? How is it obtained? By chemically processing calcareous rocks.

Form: Powder.

Additional information:

- Also referred to as sodium hydrogencarbonate, sodium bicarbonate, or baking soda.
- It is an acidity controller and neutralizer of acids. Its use as a chemical leavener, together with an acid, stems from the fact that when combined it reacts to form carbon dioxide gas.
- It should be noted that dissolved bicarbonate gives an alkaline pH that can change the texture of vegetables and may alter their organoleptic properties. This can be neutralized at the end of the cooking process by adding acid. It avoids the decomposition of chlorophyll.
- Due to its sodium content, it may increase the salty flavor of food.

General uses:

- **In the retail food industry:** Acidity regulator of some prepared products. Chemical leavener in baked goods.
- **In restaurants:** Chemical leavener used to reduce the calcium in water and thereby improve the conditions for cooking vegetables. It avoids the loss of the green color (chlorophyll) of the vegetables when cooked, nevertheless it can have collateral effects (vitamin destruction, disagreeable flavor, etc.)

Quantity and instructions for use:

- **Maximum/minimum quantity:** QS (minimum needed to obtain the desired effect).
- **Basic quantity for cooking:** QS.
- **Instructions for use:** Add either directly or dissolved to the food or preparation.

Biochemical compound
scientific concepts

What is it? A type of chemical compound produced by living beings.

Additional information:

Made up of molecules with carbon, hydrogen, and oxygen (for example, carbohydrates) and also nitrogen and sulfur (for example, proteins). Other elements are present in certain compounds (iron, copper, etc.), but are not widespread.

Biochemistry
scientific concepts

What is it? A science that studies all chemical phenomena related to living organisms.

Biodegradable product
scientific concepts

What is it? A product that decomposes over time due to the action of microorganisms. All food products are biodegradable, except for salt, water, and some additives. There is a push in the retail food industry to increase the use of biodegradable packaging in place of the traditional non-biodegradable packaging.

Biological process
scientific concepts

What is it? A physicochemical transformation involving biological substances or microorganisms. Examples: yogurt preparation, bread fermentation, milk curdling, etc.

Biology
scientific concepts

What is it? A science that studies all phenomena related to living organisms.

Biotechnology
scientific concepts

What is it? A technology that uses living organisms (cells, molecules, live tissues) to produce medicines, foods, or other products, or that modifies the genetic characteristics of a species or individual to make it more resistant to a disease, more productive, etc.

Additional information:

Production of bread, beer, wine, cheese, etc. is an example of traditional biotechnology; whereas the genomic manipulation of microorganisms, animals, and plants to achieve more resistance to diseases and more production are examples of modern biotechnology.

Bitter
organoleptic perceptions

What is it? One of the basic tastes.

Additional information:

- In the retail food industry quinine is considered to be the standard of bitterness. However, caffeine is increasingly being used as a reference.
- Other bitter products: gentian, caffeine.
- It is a taste that is readily rejected in childhood, as it is instinctively associated with venomous substances; later it becomes part of people's eating routines out of habit.

Bleach
mineral products

What is it? A liquid mixture whose main component is sodium hypochlorite, which prevents the proliferation of microorganisms in foodstuffs and consequently prevents possible infections.

Additional information:

- If an eggshell is rinsed in water with a few drops of bleach, salmonellosis can be avoided.
- If vegetables are washed in water with a few drops of bleach, microorganism infections can be prevented.
- It has traditionally been used to make water drinkable.

- Commercial bleaches exist that are designed especially for foodstuffs.

General uses:

- **In the retail food industry:** Disinfectant.
- **In restaurants:** Disinfectant.

Quantity and instructions for use:

- **Maximum/minimum quantity:** QS (minimum required to obtain the desired effect).
- **Basic quantity for cooking:** 2–3 mL per liter of water. Only as a disinfectant, especially for vegetables, fruit, and eggs.
- **Instructions for use:** Rinse the foodstuff to be disinfected in the water-bleach solution. Then rinse with water.

Bloom values

food concepts

What are they? A measure of the gelling strength of a product. A device called a bloom gelometer is used. It is mostly used for gelatins.

Additional information:

- Resistance is determined by the weight in grams required to oppose gel with a deformation of 4 mm in a watery solution with 12.5% gelatin.
- Gelatins can be identified with varying strengths, with bloom values of between approximately 75 and 300. The higher the bloom number, the stronger the gel produced. The gelatin (fish gelatin) normally used in the form of translucent leaves has a bloom value of 220.

Boiling (temperature)

See **Boiling point**

Boiling point

scientific concepts

What is it? The temperature at which a substance transforms from liquid to steam or vice versa.

Additional information:

- It is defined as the point at which a liquid begins to boil, and depends on external pressure.
- It is also referred to as the steam or vaporization point. For example, the boiling point of distilled water (with no other dissolved component) at sea-level atmospheric pressure is 100°C. If salt, sugar, or other components are dissolved in water, the boiling point rises in accordance with the concentration of the components.
- The higher the pressure, the higher the boiling point. In pressure cookers, water boiling temperatures can reach up to 130°C.
- When pressure is reduced with a vacuum pump, or at high altitudes, the boiling point decreases. For example, at one tenth of an atmosphere (0.1 atm), the boiling temperature of water is 40–50°C. Another example is ethyl alcohol, which boils at 78.5°C (at atmospheric pressure).

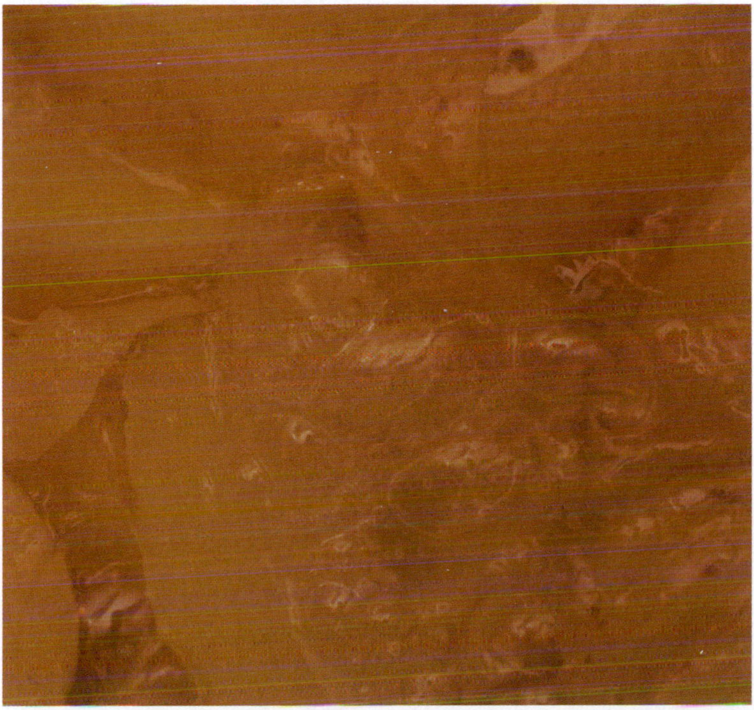

Bonding
scientific concepts

What is it?

- The linking of atoms to form molecules. The bonding of two hydrogen atoms and one oxygen atom produces water.
- The linking of two simple molecules to form a different molecule. For example, the bonding of glucose and fructose produces sugar (sucrose).

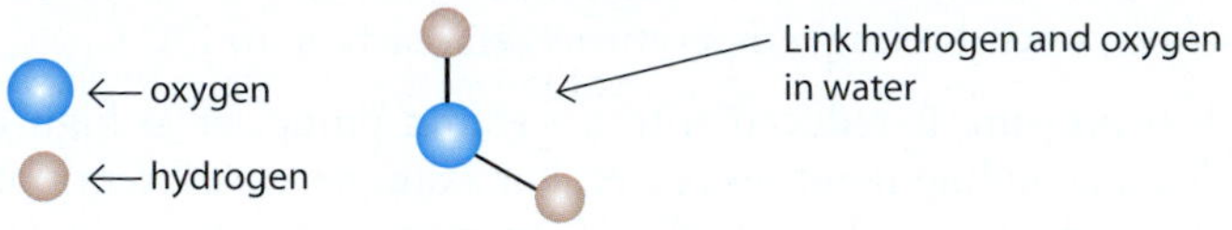
oxygen
hydrogen
Link hydrogen and oxygen in water

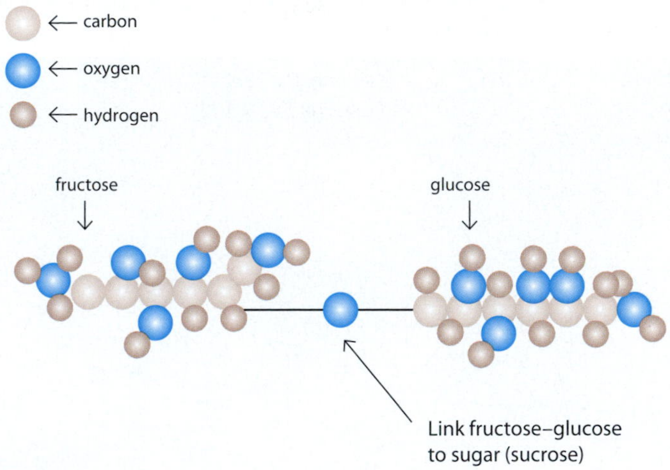
carbon
oxygen
hydrogen
fructose
glucose
Link fructose–glucose to sugar (sucrose)

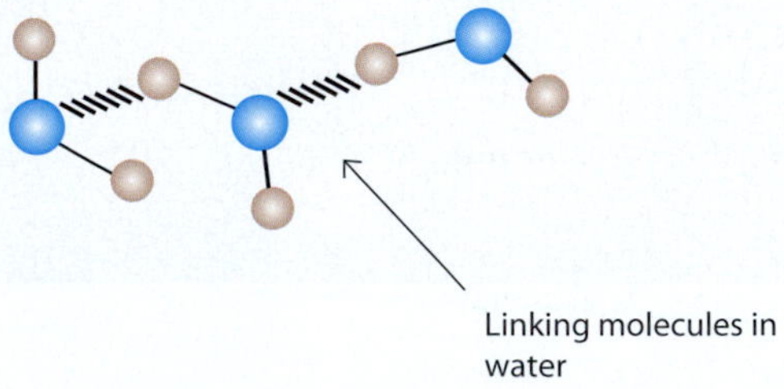
Linking molecules in water

- The linking of molecules with each other (intermolecular bonding). This bonding does not give rise to molecules that are different from the initial molecule. For example, it confers its structure to proteins. In the case of water, molecule bonding is responsible for its high fusion and boiling points.

Brix (values)
food concepts

What are they? A measure of the quantity of sugar in blended food.

Additional information:

- Normally expressed in percent and equal to the grams of sugar in 100 g of the product.
- Also used for products that are not defined as blended. For example, in commercial glucose (glucose syrup), Brix values are between 80% and 90%.
- One of the factors that determine the action of the gelling agent pectin HM is that the food to be used must have a minimum Brix value of 60%.

Bromelain
See **Enzymes**

Bronze
See **Copper**

Bulking agent
food concepts

What is it? An element that gives volume or weight to a preparation.

Additional information:

- It is also known as a bodying agent.

- If the same volume effect of sugar (sucrose) is to be achieved when using aspartame (200 times sweeter than sugar) so that a spoonful of the latter has the same volume as a spoonful of the former, a product, such as dextrose or maltodextrin, must be added to make up the volume.

General uses:

- **In the retail food industry:** To increase volume in sweeteners, meat industry, pastries, etc.
- **In restaurants:** No known use.

Butyric acid

food composition—acids

What is it? An acid normally considered a fatty acid and related to the organoleptic properties of dairy products, especially cheese.

Where does it come from? / How is it obtained? It is extracted from cheese.

Form: Liquid.

Additional information:

- Also referred to as butanoic acid.
- It is formed when butter and other dairy products become rancid.
- Associated with the organoleptic properties of some cheeses.

General uses:

- **In the retail food industry:** To give the organoleptic properties of cheese to products such as potatoes and coated food.
- **In restaurants:** No known use.

C

Caffeine
food composition—alkaloids

What is it? An alkaloid contained in some foodstuffs (principally coffee or tea) that has stimulating effects.

Where does it come from? How is it obtained? It is extracted from coffee beans or tea leaves.

Form: Powder.

Additional information:

- It stimulates the nervous system.
- It has a bitter taste.
- It is found in coffee with a concentration of 1% to 2%, depending on coffee type.

General uses:

- **In the retail food industry:** Colas and other "energy" drinks.
- **In restaurants:** No known use.

Calcium
food composition—minerals

What is it? A metallic chemical element that is a component of certain mineral salts.

Additional information:

- It is a component that is indispensable for living organisms, fundamental for the constitution of the organism, especially for bones and teeth.

It is consumed mainly through food and, to a lesser degree, through water. Some foodstuffs with high levels of calcium are dairy products (milk, 0.12%), nuts (almonds, 0.25%) and vegetables (spinach, 0.1%). Parmesan cheese is the foodstuff that contains the most calcium, with 1.275%.

Calcium chloride

See **Calcium salts**

Calcium gluconate

See **Calcium salts**

Calcium gluconolactate

See **Calcium salts**

Calcium lactate

See **Calcium salts**

Calcium oxide

See **Lime**

Calcium salts

food composition—minerals

What are they? Salts formed by calcium and other components.

Where do they come from? How are they obtained? They are extracted from dairy and mineral products, etc.

Form: Granulated or in powder in calcium chloride, calcium lactate, etc. They are also found in water solutions.

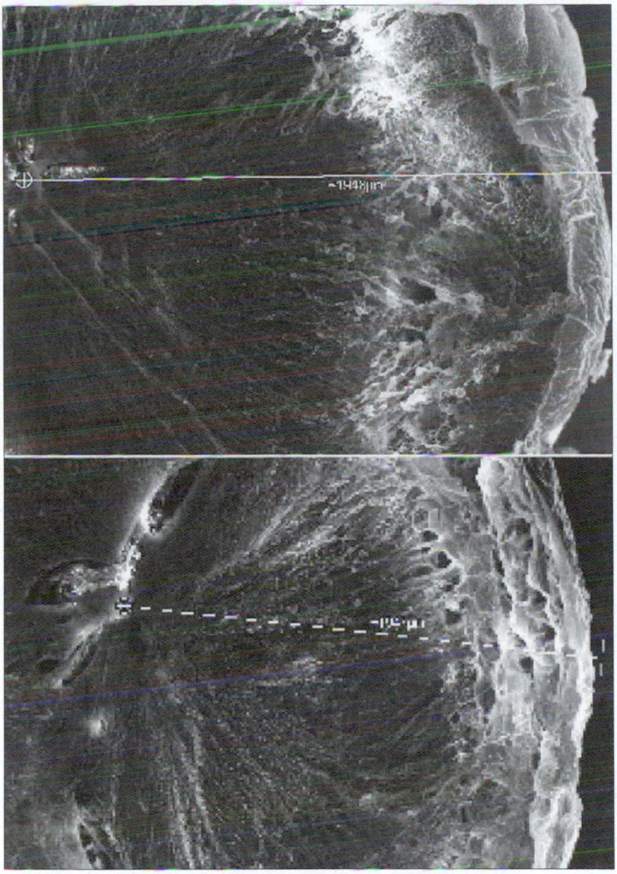

The top photograph shows an alginate sphere that has been submerged for 1 minute in the calcium chloride base; the bottom image is of a sphere that has been left for 5 minutes in the same base. Comparison of the two photographs reveals the greater penetration of calcium in the second case. After a few more minutes, the spherification would be complete. (Photographs taken with an electron microscope by Fernando Sapiña and Eduardo Tamayo from the University of Valencia.)

Additional information:

- Some gelling agents (alginates) need calcium to be effective. This property has been used in cooking to produce controlled external jellification called "spherification."
- Calcium chloride is normally used in spherification operations, although the suitability of other salts, such as calcium lactate, calcium gluconolactate, and calcium gluconate is being researched.

General uses:

- **In the retail food industry:** It is used to increase the proportion of calcium in some foodstuffs.
- **In restaurants:** In spherification (in the form of calcium salts, preferably calcium chloride).

Quantity and instructions for use:

- **Maximum/minimum quantity:** QS (minimum needed to obtain the desired effect).
- **Basic quantity for cooking (for calcium chloride):** For basic spherification, 0.5–1%; for inverse spherification, concentrations are in the experimental phase.
- **Instructions for use:** For basic spherification, mix calcium chloride with water first; for inverse spherification, add calcium chloride directly into the product (in the experimental phase).

Caloric sweetener

food concepts

What is it? A name given to a series of additives that are used as low-intensity sweeteners.

Additional information:

- Also known as nutritional sweeteners.
- All products referred to under this name belong to the polyol chemical group. *See* Polyols.
- Caloric sweeteners are based on the following properties:
  - A mild sweetness that is equal to or less than sugar.
  - Sensation of freshness in most cases.
  - They are laxative at doses greater than 60 g/kg.
  - Some are hygroscopic (they absorb water).
  - They give a certain texture to products.
- They are catalogued as additives.

Calorie
scientific concepts

What is it?

- **Scientifically:** An energy unit. It is defined as the energy required for 1 gram of water to increase its temperature by 1°C. The joule (J) that equals 0.24 calories is currently more widely used. Multiples of this unit are kcal and kJ (1 kJ = 0.24 kcal).
- **Dietetically:** The term used to describe the energy provided by nutrients. The energy provided by the food we eat is expressed in kilocalories that are often understood to be calories.

Additional information:

- All food products are associated with energy; this is apparent after digestion and varies according to the nutrients.

1 g carbohydrates	4 kcal
1 g proteins	4 kcal
1 g lipids	9 kcal
1 g alcohol	7 kcal

- **For example:** Cow's milk has 67 kcal (280 kJ) per 100 g. Low-calorie diets encourage the consumption of food with low energy content. Water has no energy contribution.

Caramel (additive) (E-150)
additives—coloring agents

What is it? An ensemble of polymer-type chemical compounds (large molecules) that is used as a coloring additive (brown color).

Where does it come from? How is it obtained? From the Maillard reaction that takes place when sugars are heated, normally in the presence of acids or bases.

Form: Liquid or powder.

Additional information:

There are many types of caramel that vary in accordance with the preparation method used.

General uses:

- **In the retail food industry:** Drinks (colas, beer, alcohol), baking, sweets, meats, ice creams, yogurt, soups, etc.
- **In restaurants:** Baking, sweets, meats.

Carbohydrates

food composition—carbohydrates

What are they? Biochemical compounds that provide the organism with energy or fiber.

Additional information:

- Also referred to as glucids.
- They are composed of carbon, hydrogen, and oxygen.
- The simple and double carbohydrate subgroups (monosaccharides and disaccharides) give a sweet taste and are known as sugars.
- The most complex carbohydrates can be classified into:
  - Digestible but without a sweet taste (starches).
  - Non-digestible and also without a sweet taste (fiber, cellulose).

Carbohydrates		
Simple (monosaccharides)		
Double (disaccharides)		
Triple (trisaccharides), quadruple, etc.		
Complex (polysaccharides)		
Digestible Fibers	Insoluble in water	Cellulose Lignin
	Soluble in water	Gums Pectins

Carbon dioxide (E-290)

additives—preservatives
additives—gases

What is it? An inorganic chemical compound (CO_2) that is used as a preservative and gasifying additive

Where does it come from? How is it obtained? From reactions between acids and bicarbonates or carbonates, from fermentations or combustions.

Form:

- Compressed gas in capsules.
- Solid in isolated containers (at –78°C at atmospheric pressure).

Additional information:

- It is also referred to as carbonic anhydride.
- It is one of the gases responsible for the greenhouse effect (progressive heating of the Earth).
- It is referred to as dry ice in solid state, at an approximate temperature of –78°C. Above this temperature it is a gas; in fact, it converts from solid to gas directly without converting to liquid. This phenomenon, called sublimation, has been used to produce smoke effects in entertainment.
- It is the gaseous component of fizzy drinks.
- Gastronomically, it is responsible for most gasification processes in the retail food industry.

General uses:

- **In the retail food industry:** Used in the MAP (Modified Atmosphere Packaging) technique in order to improve the preservation of some foodstuffs in vacuum. In carbonic drinks.
- **In restaurants:**
  - In capsules for soda-type siphons and as a propellant.
  - It can be used as a cooling agent because of its low temperature in solid state (dry ice); it should be kept in freezers. It has a limited duration because of its conversion to gas.

Quantity and instructions for use:

- **Maximum/minimum quantity:** QS (minimum needed to obtain the desired effect).
- **Basic quantity for cooking:**
  - Gas capsules of 10–15 mL approximately.
  - When solid (dry ice), the quantity used depends on what needs to be cooled.
- **Instructions for use:**
  - As a gas it is used in capsules in siphons.

- In solid state, direct introduction in the product or in the external cooling solution.

Carboxymethylcellulose (CMC) (E-466)

additives—thickening agents

What is it? A fibrous carbohydrate that is used as a thickening additive. It has the properties of a hydrocolloid.

Where does it come from? How is it obtained? From the reaction that takes place when carboxymethyl groups are added to plant cellulose.

Form: Powder.

Additional information:
It is also known as cellulose gum.

General uses:

- **In the retail food industry:** Soups, vinaigrettes, as a suspending agent (prevents precipitation of solids in liquids), in emulsions, etc.
- **In restaurants:** In experimentation.

Carmine

See **Cochineal**

Carob gum (E-410)

additives—thickening agents
additives—stabilizers

What is it? A fibrous carbohydrate from the galactomannan group that is used as a thickening and stabilizing agent. It has the properties of a hydrocolloid.

Where does it come from? How is it obtained? From grinding and refining carob.

Form: Powder.

Additional information:

- It is also referred to as carob bean gum.
- Very complex polysaccharide capable of producing highly viscous solutions.
- When mixed with other polysaccharides, it modulates their gelling properties. It gives elasticity to gels formed by agar-agar and carrageenans, avoiding cracking and reducing syneresis.
- It produces a gel with xanthan gum.

General uses:

- **In the retail food industry:** A suspension stabilizer in soft drinks, soups, sauces, baking, dough, jams, canned vegetables and cream.
- **In restaurants:** It is used in ice cream bases and pastry work.

Carotenes (E-160)

food composition—pigments and other compounds
additives—coloring agents

What are they? Coloring agents from the carotenoid group, mainly provitamin A (a product that becomes vitamin A), which provides an orange color.

Where do they come from? How are they obtained? Mainly from extraction of carrots. Also from other vegetables, algae, and animals. Also it can be obtained by a process of fermentation of the microorganisms (*Blakeslea trispora*).

Form: Powder or liquid in different concentrations.

Additional information:

There are various types, and the majority can be transformed into vitamin A in the intestinal wall. The carotene in carrots is called beta carotene.

General uses:

- **In the retail food industry:** Drinks, dairy products, margarines, prepackaged meals, etc.
- **In restaurants:** Indirectly through food: carrot, red pepper, citrus fruit, egg yolk, etc.

Carrageenans (carrageens) (E-407)

additives—gelling agents
additives—thickening agents
additives—stabilizers

What are they? Fibrous carbohydrates that are used as gelling, thickening, and stabilizing additives. They have the properties of hydrocolloids.

Where do they come from? How are they obtained? By physicochemical treatment of red algae from the Rhodophyta family (*Chondrus, Gigartina, Furcellaria*).

Form: Powder.

Additional information:

- The name comes from the Irish village Carragheen, where they have been used in food and medicine since the Middle Ages under the name Irish moss.
- The most important types of algae and carrageenans that are extracted are
  - *Chondrus chispus* (Irish moss): Kappa, iota, and lambda
  - *Gigartina radula*: Kappa and lambda
  - *Euchema cottoni*: Kappa and iota
  - *Furcellaria fastigiata*: Furcellaran
- Three kinds of carrageenans—kappa, iota, and lambda—are marketed. The only difference between them is the electric charge of the components that give them different properties. *See* Electric charge.
  - **Kappa** has hardly any electric charge and forms gels when the hot solution it has been dispersed in cools down. They are thermoreversible gels; they may liquidize again when reheated and gel again when cold. The gel fusion temperature is 70–80°C.
  - **Iota** has a moderate charge and forms less rigid gels. It is unusual in that, if it breaks down, it can be reconstituted at a normal temperature by simply leaving it to rest. *See* Thixotropy.
  - **Lambda**, with a strong electric charge, cannot form gels, but is a good thickening agent and suspender of solid particles.

Casein
food composition—proteins

What is it? A protein that has emulsifying, stabilizing, thickening, and gelling properties.

Where does it come from? How is it obtained? It is precipitated from milk, whether by the action of rennet at 35°C or by acids.

Form: Powder in casein and caseinate varieties.

Additional information:

- Milk has a 3.5% protein content, of which 76% is casein.
- It can be curdled by acidification or by enzymes such as rennin.

General uses:

- **In the retail food industry:** Dairy products (cottage cheese, cheese, junket, etc.). Whiteners for coffee, ice creams, creams, meringues, drinks, soups, stocks, sauces, etc.
- **In restaurants:** In cheese making and experimentation.

Cell
scientific concepts

What is it? A functional and structural basic unit of all living organisms where the essential reactions of the living being take place.

It is composed of

- **Cellular membrane.** Where interchanges of nutrients or expelled products take place. *See* Osmosis.
- **Cytoplasm.** The place where the processes of nutrition and relations that enable the survival of species to take place.
- **Nucleus.** This contains the chromosomes that contain the genes that regulate the characteristics of each species and where reproduction processes occur.

Cellulose
food composition—carbohydrates

What is it? A fibrous carbohydrate from the glycan family, a component of many vegetable products.

Where does it come from? How is it obtained? It is extracted from wood or cotton pulp.

Additional information:

- It is the main constituent of the cell walls of high-fiber vegetables and plant stalks.
- Formed by repetitions of linked glucose monomers (glucose polymer) and classified as non-soluble fibers.

Main derived additives in food:

- **Microcrystalline cellulose** (MCC) E-460. Used as a bulking agent and to make sauces, foams, and as a viscosity agent.
- **Methylcellulose** (MC) E-461. A high-temperature gelling agent used in crème caramels, béchamel sauce, pizzas, etc.
- **Carboxymethylcellulose** (CMC) E-466. A suspending agent (prevents sedimentation) in vinaigrettes, used to form films.
- Other derivatives of cellulose are hydroxypropylmethylcellulose E-464 and ethylcellulose (EC) E-463, which, like methylcellulose, are high-temperature gelling agents.

Centrifugation

See **Centrifuge.**

Centrifuge

technology—devices

What is it? A device that separates the particles of a liquid by the use of centrifugal force.

How does it work? It is based on the rotation of a series of tubes on a vertical axis. This produces a centrifugal force that increases the force of gravity and causes precipitation of the product in the tubes.

Additional information:

- Normal centrifuge increases the force of gravity approximately 1,000 times, but when working at 12,000 rpm (revolutions per minute) more than 12,000 gravities can be reached. The conversion of revolutions per minute to gravities depends on the centrifuge, basically on the rotational radius.

- In cooking, centrifuges allow different components of one foodstuff to be separated according to density.

General uses:

- **In the retail food industry:** Separation of the components of milk, residues of oils, etc.
- **In restaurants:** Little used. It has been used to separate solid components and produce clarifications, practically without heat, using both the solid concentrate at the base and the liquid; for example, tarragon concentrate.

Chemical compound

scientific concepts

What is it? A pure substance consisting of two or more different chemical elements that can be separated into simpler substances by chemical reactions. Chemical compounds have a unique and defined chemical structure; they consist of a fixed ratio of atoms that are held together in a defined spatial arrangement by chemical bonds.

Additional information:

- Today, some 27 million different compounds are recognized.
- Food products are composed of chemical compounds, either single compounds (sugar, common salt, etc.) or a mix of different compounds (milk, vegetables, etc.).

Chemical compounds can be classified into

- **Organic compounds:** Formed by carbon and hydrogen, normally constituting chains, and usually with oxygen. They may also contain other elements (nitrogen, phosphorus, etc.), for example, carbohydrates, lipids. Most food products are organic; minerals and water are the exceptions.
- **Inorganic compounds:** Formed by combinations of the elements of the periodic table, except the carbon and hydrogen groups; for example, water, salt, etc.

Chemical elements
scientific concepts

What are they? All the simple components of matter, classified in the periodic table; for example, oxygen, chlorine, carbon, etc. *See* Periodic table.

Chemical leavener
food concepts

What is it? A product with a carbonate/bicarbonate base with or without the addition of an acid. The reaction of the two produces carbonic gas (carbon dioxide).

Additional information:

- It is used to give floury doughs a springy effect (spongecakes).

Chemical process
scientific concepts

What is it? A transformation where there is a change in the compositions of the substances that make up the product.

Additional information:

- A good example is digestion: starches are transformed into simple sugars; proteins become amino acids, etc.
- When meat or fish is cooked on the barbeque, Maillard reactions occur (as do many others) and are noticeable at 130°C.
- Another example is sugar caramelization, which occurs at approximately 150°C; the white and sweet sugar disappears and is transformed into a brown bittersweet mass. If the temperature is increased further, the caramel becomes even darker and loses all of its sweet taste, and when heating is continued, a black residue that is mainly carbon is all that remains.

Chemistry

scientific concepts

What is it? A science that studies the composition of matter and its transformations on contact with various substances and the energy changes that take place within them.

Chlorine

food composition—minerals
mineral products

What is it? A chemical element that tends to be associated with other elements, mainly in salts (sodium chloride, calcium chloride, etc.).

Additional information:

- As a chloride, it is a component of different mineral salts. Vital for living organisms; the human organism contains 1.1 g per kg.
- Indispensable for equilibrium with sodium and in gastric juices. Its presence in the organism is due mainly to the digestion of common salt (sodium chloride).
- It is also used as a disinfectant, forming part of compounds such as hypochlorite (basic bleach product).

Chlorophyll (E-140)

additives—coloring agents
food composition—pigments and other compounds

What is it? A natural pigment of plants, it is a formed by a porphyrinic group with magnesium and other components. Some stabilized forms are considered as a coloring additive that is used for its green color

Where does it come from? How is it obtained? It is extracted from herbs, alfalfa, and other edible green vegetables (spinach, Swiss chard, etc.). The easiest way to extract it from plants is with ethyl alcohol, because it readily dissolves in this liquid. Later, the chlorophyll is obtained by evaporating the alcohol.

Form: Powder or in solution.

Additional information:

In an acidic medium, it degrades and allows other pigments such as carotenes (orange) that were previously hidden by the chlorophyll to be revealed. As a result, the vegetables turn a brownish color; for this reason it is essential to reduce the acidity (increase the pH) by adding, for example, bicarbonate. The color is then fixed.

General uses:

- **In the retail food industry:** Ice creams and dairy products.
- **In restaurants:** In experimentation.

Cholesterol

food composition—lipids

What is it? A simple steroid-type lipid (organic compound characterized by cyclic structures).

Additional information:

- It exists exclusively and naturally in the animal kingdom and its accumulation in the human organism causes arteriosclerosis (hardening of the arteries) with consequences such as angina or myocardial infarction.
- The reduction of cholesterol levels is associated with the consumption of products with polyunsaturated fatty acids; for example, oily fish.
- Nowadays, polyunsaturated fatty acids may be found in other foodstuffs that do not naturally contain them and are sold as anti-cholesterol products; for example, margarine with omega 3.
- Two types of proteins, fats, and cholesterol aggregates can be defined:
  - High-density cholesterol/HDL—beneficial to the organism.
  - Low-density cholesterol/LDL—considered harmful.
- It is present in abundance in egg yolks and animal fats.

Quantity of Cholesterol in Some Foodstuffs (in mg per 100 g)	
Cow's milk	13 mg
Single cream 35% fat	110 mg
Egg yolk	1,100 mg
Cod liver oil	500 mg
Butter	250 mg
Brie	100 mg
Caviar	300 mg
Oysters	260 mg
Pork brains	2,000 mg
Liver (pork, chicken, duck, etc.)	300–500 mg
Fruit, vegetables, vegetable oils or margarines and non-animal products in general	0 mg

Chromatograph

technology—devices

What is it? A device used to analyze the components of a mixture.

How does it work? It separates and identifies the components of a mixture through the effect of the different speeds at which they move inside the device.

Additional information:

There are different types of chromatographs. The gas chromatograph, which is applicable to gases and volatile substances, and liquid chromatography is applicable for liquid or dissolvable, non-volatile samples. A detector is used to identify the type and quantity of products in the sample, based on the speed of the flow in the tube.

General uses:

- **In the retail food industry:** To analyze food and carry out quality controls.
- **In restaurants:** No known use.

Chymosin (rennin)

food composition—proteins

What is it? An enzyme contained in rennet, the liquid separated by part of the stomach of ruminants. This enzyme is capable of coagulating (setting) milk or similar products.

Where does it come from? How is it obtained? From rennet.

Form: Liquid in different concentrations.

Additional information:

- This enzyme can also be found in some plants, such as thistles, and is much used in cheese preparation.

General uses:

- **In the retail food industry:** Cheese preparation.
- **In restaurants:** Cheese preparation.

Citric acid (E-330)

food composition—acids
additives—acidity regulators
additives—preservatives

What is it? An organic acid used as an acidity regulator and as a preservative.

Where does it come from? How is it obtained? It is mainly present in citrus fruit but also in strawberries, pineapples, raspberries, etc. It is extracted by physicochemical treatment of maple fermentation.

Form: Crystallized or dissolved in different concentrations.

General uses:

- **In the retail food industry:**
  - Used to acidify (pH control) and to preserve animal or vegetable oils and fats. Used as a preservative in quick-cook rice, cocoa and chocolate, milk and dairy products (pasteurized whole cream), bread, fresh pasta.
  - Prepared vegetable products: low-energy jams, jellies and marmalades, fruit-based preparations for spreading, including those with reduced energy values, juices, etc.

- **In restaurants:** Used to add or enhance the acidic taste. Used to encourage jellification processes (fruit pastes, etc.).

Quantity and instructions for use:

- **Maximum/minimum quantity:** QS (minimum needed to obtain desired effect), except cocoa and chocolate (5 g/kg) and fruit and nectars (3 g/kg).
- **Basic quantity for cooking:** Depends on the preparation.
- **Instructions for use:** It is mixed or added directly to the preparation, in powder or liquid form (dissolved in water).

Clarification

physical or chemical processes

What is it?

- **Scientifically:** Removal of cloudy particles from a liquid. There are several methods:
  - Filtration: With funnels and suitable filters; for example, wine clarification with bentonite, cellulose filters, etc.
  - Decantation: The liquid is left to rest so that the particles sink to the bottom.
  - Centrifugation: The liquid is placed in a device that increases g-forces and separates the particulate components. These will be finally separated by decantation.
- **Gastronomically:** In gastronomy, clarification is mainly considered in two cases:
  - Consommé: by coagulation of egg white albumins.
  - Butter: by heat application and decantation.

Coagulation

scientific concepts

What is it? An action whereby large conglomerations of molecules in a liquid agglutinate into the form of a gelatinous solid, which can help the separation between the aforementioned solid and the rest of the liquid.

Additional information:

Normally associated with a later phase of protein denaturing. This can be produced by heat, by cooling, by enzyme action or by adding products such as acids, alcohols, etc; for example, the coagulation of egg whites by heat, or the use of the rennin enzyme to manufacture cheese by curdling milk.

Coating agent

food concepts

What is it? A product that creates a shiny or protective layer when applied to the exterior surface of foodstuffs.

Additional information:

The best coating agents are waxes; for example, white and yellow beeswax (E-901), carnauba wax (E-903), etc.

General uses:

- **In the retail food industry:** Cheese wax, glazing for baked goods, coating on chewing gum, etc.
- **In restaurants:** Gold, silver coatings, and glazing on baked goods.

Cochineal (E-120)

additives—coloring agents

What is it? A coloring additive (red or violet) whose main active component is carminic acid.

Where does it come from? How is it obtained? It is extracted from natural products within the shell of dried females of an insect (*Coccus cacti*) that lives in cacti in the Canary Islands and South and Central America.

Form: Powder.

Additional information:

- Between 100 and 150 insects are needed to obtain a gram of cochineal.
- It was widely used in the Middle Ages for cosmetics and dying.

General uses:

- **In the retail food industry:** Sausages, shellfish, syrups, baking, chewing gum, dairy products, baked goods, drinks, etc.
- **In restaurants:** Pastry coloring.

Collagen

food composition—proteins

What is it? A protein that has emulsifying, airing, and gelling properties.

Where does it come from? How is it obtained? It is extracted from pork, veal, chicken and fish.

Additional information:

- The most abundant type of protein in superior vertebrates, for example, in mammals (20 to 25% of proteins).
- It gives rigidity to meat. This characteristic can be modified, for example, with prolonged cooking in water.
- Collagen fibers swell up in acidic and alkaline products and in the presence of some mineral salts.
- It is abundant in muscles, tendons, cartilage, and skin.

Colloid

scientific concepts

What is it? A gel or the dispersion of large molecules or aggregates, normally in a aqueous solution, where they are not dissolved although at a glance they would seem to be. The resulting mixture tends to be notably more viscous than water.

Additional information:

- All hydrocolloids form colloidal solutions.
- Colloid solutions are also referred to as colloidal dispersion or emulsion.
- A culinary example is when water is mixed with oils or fats in general to form mayonnaise, vinaigrettes, etc.

Coloring agent
scientific concepts

What is it? A substance capable of conferring a determined color to others, either because it possesses the color or because it can produce it in certain conditions.

Additional information:

- Normally, food coloring agents are artificial additives (tartrazine, Ponceau 4R, etc.) or natural (cochineal, curcuma, chlorophyll, etc.). The latter may also be obtained by chemical synthesis.
- Coloring agents can be soluble in water (hydrophile) and soluble in fats (lipophile).
- All coloring agents may be affected by light, temperature, acidity, oxygen, the presence of microbes, etc., leading to changes in or disappearance of the color tone.

Copper (E-171, E-172)
additives—coloring agents

What is it? A mixture of surface coloring additives formed basically by titanium dioxide (E-171) and iron oxides (E-172), which gives a range of colors that may be likened to copper or bronze.

Form: Powder.

General uses:

- **In the retail food industry:** In baking and confectionery. E-172 is used as a surface coloring agent for cheeses.
- **In restaurants:** Coatings, on rare occasions.

Quantity and instructions for use:

- **Basic quantity for cooking:** QS (minimum needed to obtain the desired effect) for the coating of preparations.
- **Instructions for use:** Brush the surface of the preparation with the coating. Mixing with water speeds up the process.

Color	Natural Coloring Agents	Artificial Coloring Agents
	Calcium carbonates Titanium dioxide	
	Curcumin Riboflavin	Tartrazine
	Gold	
	Carotenes Anthcyanins (blue-red, orange) Xanthophyll (yellow to red)	Quinoline yellow Quinoline orange S
	Caramels Iron oxides or hydrides	Brown FK Brown HT
	Anatto, bixin or norbixin Pepper, capsanthin and capsorubin extract	
	Cochineal or cochineal carmine Lycopene Betanine or beet red	Red cochineal A and Ponceau 4R Azorubine Amaranth Red 2G Allura red AC Lithol rubine BK
		Erythrosine
		Patent blue V Indigotine and indigo carmine Brilliant blue FCF
	Chlorophylls	Green S
	Aluminum Silver	
	Carbon	
	Melanin or squid and cuttlefish ink, etc.	Black PN and Brilliant black BN

In cooking and in the retail food industry different tones may be added to food-stuffs thanks to natural and artificial coloring agents.

Curdlan

additives—gelling agents
additives—thickening agents

What is it? A fibrous carbohydrate that is used as a gelling and thickening additive. It has the properties of a hydrocolloid.

Where does it come from? How is it obtained? It is extracted from the bacterium *Agrobacterium* biovar 1 *(Alcaligenes faecalis).*

Form: Powder.

Additional information:

- It is not catalogued in the European Union and therefore does not have an E number. It is not authorized as an additive for industrial application, but it is used in other countries, such as Japan and the United States.
- It can be used in cooking (although it must be bought in the countries where it is consumed).
- It has a characteristic that differentiates it from others as a gelling agent: when hydrated at between 60 and 80°C a type of gel is produced that then remains when cooling.

General uses:

- **In the retail food industry:** Outside of the European Union, it is used in surimi, pasta, prepackaged meals, meat products, ice creams, and as a soluble fiber in dietetic formulas, etc.
- **In restaurants:** In experimentation.

Cyclamate (E-952)

additives—sweeteners

What is it? An organic salt and artificial additive that is used as a sweetener.

Where does it come from? How is it obtained? Synthesis in the chemical industry; it is obtained from petroleum derivates (cyclohexylamine and chlorosulfonic acid).

Form: Powder.

Additional information:

- Full name: Sodium cyclamate.
- Sweetening power: 40 times the sweetness of sugar (sucrose).
- It has been questioned for health reasons.

General uses:

- **In the retail food industry:** Soft drinks, sugar substitute for diabetics and in the confectionery industry (sweets, chewing gum, jelly candies, etc.).
- **In restaurants:** No known use.

D

Decantation

physical or chemical processes

What is it? A separation method for components when at least one of which is liquid. It is based on the density differences between undissolved products.

Additional information:

The densest product sinks to the bottom and the least dense remains on the surface.

Decantation funnel

technology—devices

What is it? A utensil that separates the immiscible components of liquid mixtures, for example, oil and water.

How does it work? The mixture is placed in a container with a stopcock and when the phases have been differentiated, the stopcock is opened in order to separate the components.

Examples of application: Used to separate flavors according to whether they are soluble in water or in fat.

Dehydration

physical or chemical processes

What is it? The partial or total extraction of water from a foodstuff in order to preserve the product or to obtain a lighter product with new forms, textures, etc.

Denatured proteins

physical or chemical processes

What are they? Proteins that have lost their tridimensional structure and have consequently acquired different properties.

Additional information:

- Denatured proteins are responsible for different cooking processes; for example, cottage cheese or junket are produced when casein is denatured by the action of an acid or an enzyme.
- Proteins are denatured by physical processes (the action of heat, cold, mechanical actions, etc.) or chemical processes (addition of acids, alkalis, alcohols, enzymes, etc.).
- It is an irreversible process, i.e., the original proteins cannot be restored.

Density

scientific concepts

What is it? The ratio between the mass and the volume of a substance.

Additional information:

- Densitometers are used to measure the density of liquids.
- Density is used in the retail food industry to measure alcohol in wine, beer, distilled beverages, etc. It is also measured to analyze fatty materials in a preparation (for example, the density of oil used for frying indicates its possible toxicity), also salt in brine, sugars in jams, etc.

Examples of densities:

- Water 1 g/mL (1 kg/L). 1 liter of water weighs 1 kg.
- Olive oil 0.83 g/mL (0.83 kg/L). 1 liter of oil weighs 0.83 kg.
- Ethyl alcohol 0.79 g/mL (0.79 kg/L). 1 liter of ethyl alcohol weighs 0.79 kg.

Dextrin

food composition—carbohydrates

What is it? A carbohydrate derived from starch that is used as a sugar supplement.

Where does it come from? How is it obtained? When starch is broken down.

Form: Powder.

General uses:

- **In the retail food industry:** A water-soluble glue in sweet foods and a fat substitute in dietetic products.
- **In restaurants:** No known use.

Dextrose

food composition—carbohydrates

What is it? In the world of cooking and the retail food industry, it is the term normally used when referring to glucose. *See* Glucose.

Dextrose equivalent (DE)

food concepts

What is it? It is the indication of how much dextrose (glucose) or equivalent sugars (fructose, galactose, etc.) are contained in a mix of carbohydrates.

Additional information:

- It is a term frequently used in glucose syrups. For example, a syrup with a DE of 20 has 20% dextrose mixed with other carbohydrates. *See* Glucose syrup.

Diatoms

mineral products

What are they? Very porous minerals used to clarify wine.

Where do they come from? How are they obtained? By physicochemical treatment of fossil remains of unicellular sea algae.

Form: Powder.

Additional information:

- They are the remains, accumulated over centuries, of unicellular marine algae (tripoli, diatomaceous earth).

- They are very absorbent, which is why they are used in the wine industry as clarifiers to eliminate floating solids in wine. Recently, precision filtrations have led to their being used less frequently.

General uses:

- **In the retail food industry:** Wine clarification.
- **In restaurants:** No known use.

Dietetics

scientific concepts

What is it? The area of nutritional science that investigates food consumption, according to the conditions of a person (sex, age, work, physical activity, illnesses, etc.) to achieve a diet that helps to maintain an optimum state of health or that is compatible with medical treatment that affects the metabolism, bearing in mind biochemical and nutritional principles.

Disaccharide

food composition—carbohydrates

What is it? A carbohydrate formed by two molecules of simple carbohydrates (monosaccharides).

Additional information:

- The most important disaccharide is common sugar, or sucrose, which forms a daily part of the lives of the majority of the population. It is formed by bonding glucose and fructose molecules.
- It has the following properties:
  - Highy soluble in water.
  - Sweet. Sucrose (common sugar) is used as a measurement standard of the sweetening power of a product and is assigned the value 1. Fructose is between 1.1 and 1.7 and glucose between 0.5 and 0.8. These variations depend on whether the product is dissolved or crystalline, on the characteristics of the foodstuff it is applied to, and on subjectivity regarding the sweetening property.
  - It acts as a preservative when sugar proportions are high. It should constitute more than 60% of the product in order to act as a preservative and prevent the microbial effect in jams, confit fruit and other sweet foods.

- These properties, added to the fact that it is easy to obtain and apply, make it a widely used product, basically as a sweetener.
- Another important disaccharide is lactose, which is the sugar in milk (4.8%). It has low sweetening power and is formed by glucose and galactose monomers

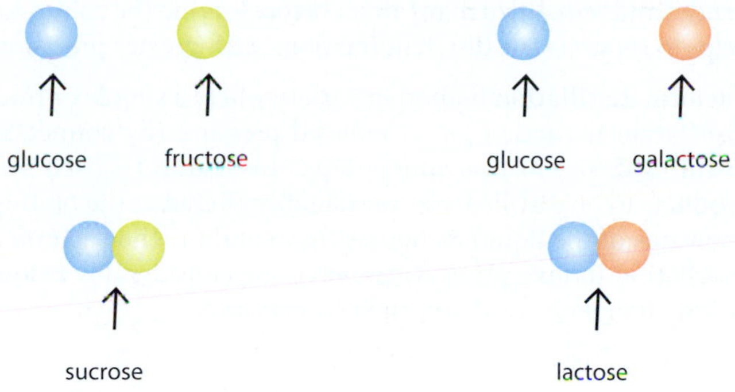
glucose
fructose
glucose
galactose
sucrose
lactose

Distillation
physical or chemical processes

What is it? A method that is based on the different volatilities of substances and used to separate soluble or miscible components.

Additional information:

It is normally used to separate the non-volatile impurities of a liquid or to separate two or more liquids that have substantially different boiling temperatures.

Distillation technique:

product to be distilled + heat action

↓

volatilization of volatile components

↓

refrigeration and steam condensation

↓

distilled product

Types of distillation:

- **Simple distillation:** Used to separate a substantially more volatile component from a mixture of components. The initial separation is not very precise.
- **Fractional distillation:** Used to separate a mixture of components with diverse volatilities. In the distillation column, the steam is condensed and redistilled many times before leaving the column, which helps to separate the different fractions with greater precision.
- **Vacuum distillation** is another variety where a simple or fractional distillation is carried out at reduced pressure (by connecting the distilling device to a vacuum pump). This method is used when the products to be distilled are considerably altered at the boiling temperatures of the liquid or liquids they contain. The vacuum allows distillation to take place at temperatures considerably below their boiling temperature at atmospheric pressure.

Dominant liquid
food concepts

What is it? It is the liquid in which a product is immersed for certain preparations (stocks, syrups, marinades, etc.). Its presence encourages osmosis, preservation, etc.

Dry ice

See **Carbon dioxide**.

Dry weight
scientific concepts

What is it? The quantity of a product that remains after complete dehydration.

Additional information: For example, in 100 g of carrot the dry extract will be

Proteins	Lipids (fats)	Carbohydrates (sugars)	Mineral Salts and Vitamins	TOTAL Dry Weight
0.8 g	0.2 g	9.2 g	1.7g	11.9 g

The remaining 88.1 g of carrots is water.

E

Electric charge
scientific concepts

What is it? A shortfall or surplus of electrons in a substance. The charges are referred to as positive and negative, respectively.

Additional information:

- Products are formed by molecules that are electrically neutral because their positive and negative components are in equilibrium.
- Due to product restructuring or to reactions that have taken place in a product, there may be parts of molecules that are not balanced and therefore have an electric charge; they are ions that can be described as electrically charged parts of a molecule. Cations have a positive charge, anions have a negative charge.
- Examples: salt is made up of chlorine and sodium atoms. Chlorine has a negative electric charge and is a chlorine ion; sodium has a positive electric charge and is denominated a sodium ion.
- Mineral salts in food compositions (iron, calcium, sodium, etc.) are really ions. The neutral or ionic (electric charge) nature of the food component results in different properties.

Electron
scientific concepts

What is it? A component of an atom with a negative electric charge (the proton is responsible for the positive charge).

Emulsifier
scientific concepts
food concepts

What is it? A product that enables the formation and maintenance of an emulsion or homogenous mixture of two immiscible liquids such as oil and water.

Additional information:

An emulsifier is a substance whose molecule is part soluble in water and part soluble in oil. This means that it is on the border of water–oil separation, guiding each part toward its accustomed phase and lowering the surface tension. This stabilizes the emulsion.

Most Common Emulsifiers	
Lipids and derivatives (animal and vegetable)	Propylenglycol esters
	Sorbitan esters
	Polyglycerol esters
	Sucroglycerides
	Sucrose esters
	Mono and diglycerides
	Fatty acid salts
	Lecithins
Carbohydrates	Starch derivatives
Proteins	Animal and vegetable

Emulsion
physical or chemical processes

What is it? A colloidal dispersion of two immiscible liquids; for example, fat and water in milk and mayonnaise are drops of oil dispersed in water, etc.

Additional information:

There are two distinct phases in an emulsion. In an emulsion one liquid (the dispersed phase) is dispersed in the other (the continuous phase). Depending on whether the internal drops are oil or water, we have the two types of emulsion:

- Oil in water (O/W) when the dispersion medium is water; for example, cream, mayonnaise.
- Water in oil (W/O) when the dispersion medium is oil; for example, margarine, ice creams.

A sensation of fattiness is noted on the palate when the outer phase (the first to come into contact with the mouth) is oil.

"Oil in water" emulsions are more easily contaminated by microbes. On the other hand, "water in oil" emulsions do not need preservatives, as microbes cannot penetrate the fatty layer to reach the water, which is the only place where they can develop. The stability of an emulsion is guaranteed by emulsifiers. *See* Emulsifier.

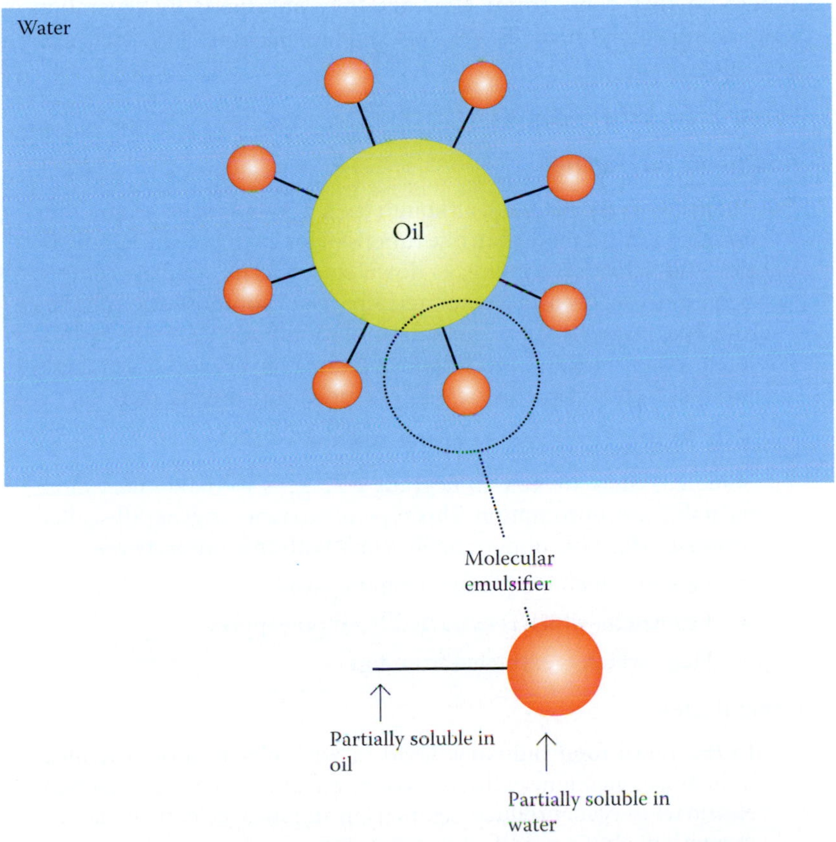
Water
Oil
Molecular emulsifier
Partially soluble in oil
Partially soluble in water

Encapsulation
physical or chemical processes

What is it? A method of protecting very delicate substances (vitamins, aromas) by enclosing them in an edible wrapping made from proteins, lipids, or carbohydrates, so that they are isolated until consumed or used.

Enzymes
food composition—proteins

What are they? Proteins that act as biological catalysts, i.e., capable of decomposing or synthesizing other substances without being affected themselves.

Where do they come from? How are they obtained? By separation of these components in foodstuffs (pineapples, papayas, figs, etc.) and in microorganisms.

Form: Granulated or powder.

Additional information:

- There are many enzymes, although their action is specific and they cannot perform more than one function; for example, during digestion, the saliva ptyalin breaks down starch to its simplest component, glucose; then proteases and lipases decompose the proteins and fats, respectively, to make assimilation possible. Once assimilated, other enzymes are responsible for converting these simple products into organic tissue, such as meat, skin, bones, etc.
- They become inactive at temperatures above 70°C.
- Some enzymes are able to degrade collagens to make meat more palatable for consumption. This type of enzymes is generally called protease. The most important enzymes with this property are
    - Papain, which is extracted from papayas.
    - Bromelain, which is extracted from pineapples.
    - Ficin, which is extracted from figs.

General uses:

- **In the retail food industry:** Used in wine clarification, the meat industry, confectionery, juices, syrups, etc.; also used to peel, soften, eliminate oxygen, produce sugars from starch, accelerate bread fermentation, obtain modified starches, etc.
- **In restaurants:** Meat glue, protein cross-linking, marinades.

Essence
food concepts

What is it? An aromatic product that is responsible for the aroma of plants and that is extracted from a plant by water distillation (carried by steam) or infusion.

Additional information:

- It is used in perfumes, medicine and also in cooking.
- An important group within the essences is the essential oils, for example, lemon oil.
- Some essences are synthetic, created to emulate the original ones, although essences are conceptually and, by definition, natural.
- Not all essences are edible.

Essential oils
See **Essence**

Ethanol
See **Ethyl alcohol**

Ethyl alcohol
food composition—alcohols

What is it? An organic product from the alcohol group. It is often referred to as ethanol, particularly in the scientific community.

Where does it come from? How is it obtained? It is obtained by distilling fermented sugar.

Form: Liquid product.

Additional information:

The ethyl alcohol (ethanol) content in beverages is measured in degrees (European degrees) that indicate the percentage of alcohol in the liquid volume. For example, a wine with 12% volume contains 12 ml of alcohol per 100 ml of liquid. The American scale uses degrees proof, which are half the value of European degrees. For example: a beverage with 50° proof has an alcoholic content equal to a beverage with 25° in Europe.

General uses:

- **In the retail food industry:** Alcoholic beverages.
- **In restaurants:** Present in alcoholic beverages used to make sorbets, gelatines, ceviches and other preparations. Its antifreeze properties are also used.

Extraction

physical or chemical processes

What is it? A procedure during which components are removed from a complex product due to the addition of a third substance.

Types and applications:

- **Liquid–liquid extraction:** The separation of two liquid components dissolved in each other. A third component that is soluble in one of the two components and immiscible with the other is added. For example: acid can be extracted from vinegar (acetic acid) by adding ether, which is not soluble in water but highly soluble in acetic acid.
- **Solid–liquid extraction:** The separation of one of the components of a mixture (one of which is a solid). For example: if water is added to marine salt (sand + salt), the sand is separated from the salt, which remains in the solution.
- Another very important example is when refined oils are obtained. When a fatty solvent is added to ground seeds, the oil in the seeds dissolves in the solvent and the oilless remains of the seeds are removed. Then the solvent is evaporated so that it separates from the oil.

Extrusion

physical or chemical processes

What is it? A procedure where a doughy mass is passed through openings at a particular temperature and pressure in order to achieve a specific shape.

Additional information:

- The application leads to totally new products with airy and crispy textures: puffed cereal, flakes, cheese curls, etc.

- Some examples of extrusion that are used frequently by chefs: pasta (macaroni), churros, etc.
- Widely used in the plastics industry.

Exudation

physical or chemical processes

What is it? A procedure during which a resin is obtained by incision in the bark of certain trees.

For example: Arabic, tragacanth, and karaya gums, which are used in the retail food industry as thickening agents, and maple sap used for making syrup are extracted by exudation.

F

Fat bloom

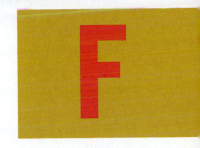

physical or chemical processes

What is it? A phenomenon that causes chocolate to be whitened by the separation of fats and their migration to the outer part of the chocolate.

Additional information:

- This whitening occurs over time and is accelerated by sudden temperature changes.
- Fat bloom can be minimized by the addition of emulsifiers other than lecithin, which is already present in many chocolates; for example, sorbitan esters, sucrose esters, etc.

Fats

food composition—lipids
food concepts

What are they?

- **Scientifically:** Fats is the generic name for organic substances (lipids), formed by the combination of fatty acids with glycerine.
- **Gastronomically:** They are substances that contain a large proportion of lipids and are used as ingredients in the kitchen, as cooking elements, as means of preservation and also as a foodstuff.

Additional information:

- Lipids, which are solids or pastes at room temperature, tend to be called fats, unlike oils that are liquid at room temperature; for example, lard, butter. *See* Lipids.

Fatty acids

food composition—acids

What are they? Organic acids (i.e., formed by carbon, hydrogen and oxygen) that tend to have a relatively high number of carbon atoms (usually 14, 16 or 18).

Where do they come from? How are they obtained? They are found in lipids.

Form: Powder, paste, or liquid.

Additional information:

- Usually found combined with glycerine, normally to form glyceride-type lipids: monoglycerides, diglycerides, and above all, triglycerides.
- If they have less than 10 carbons they have an unpleasant smell and taste on their own, but in the form of glycerides these characteristics vary substantially. A fatty or oily lipid will become rancid if the fatty acids separate (by hydrolysis of the fat or the oil).
- In the case of oils, the quantity of free acid defines the acidity degree; for example, 0.4° oil indicates 0.4% free acids (0.4 g of acid per each 100 g of oil).
- Fatty acids and other substances that may be damaging to health are liberated with prolonged use of oil or fat at high temperatures.
- Fatty acids are also present in some emulsifying additives; for example, E-471, monoglycerides, and diglycerides.

They are classified into:

- **Saturated fatty acids:** Without a double-bonded molecule, and associated with cardiovascular problems in medicine. They have high fusion points and are therefore normally found in a solid state at room temperature.
  - **Examples of saturated fatty oils with their fusion points:**

    Palmitic: 63.1°C

    Stearic: 69.6°C

    They are present in butter, beef tallow, lard, cocoa butter, palm oil, coconut oil, etc.
- **Unsaturated fatty acids:** Their fusion points are near or lower than room temperature (20–25°C). They are
  - **Monounsaturated acids:** With a double-bonded molecule and without the aforementioned health problems.

- **Polyunsaturated acids:** With more than one double-bonded molecule and associated with prevention of cardiovascular risks in medicine.
- **Examples of unsaturated fatty oils with their fusion points:**

  Oleic: 13.4°C (monounsaturated)

  Linoleic: –5°C (polyunsaturated)

  These are mostly present in olive, sunflower, peanut, and other oils.

General uses (saturated and unsaturated fatty oils):

- **In the retail food industry:** As raw materials for preparing emulsifiers.
- **In restaurants:** No known use.

Fecula

food composition—carbohydrates

What is it? The name used to denote tuber starches (potato, tapioca, etc.).

Fermentation

physical or chemical processes

What is it? A chemical transformation carried out by microorganisms (bacteria, fungi) normally combined with carbohydrates to produce alcohols or acids. There are exceptions, such as the malolactic fermentation of wine.

Additional information:

The substances produced by these microorganisms, which cause fermentation, are referred to as ferments (enzymes).

The most important fermentations are

- Alcoholic fermentation: Wine, beer, distilled beverages, etc.
- Lactic fermentation: Yogurt, kefir, cheeses, cured sausages, etc.
- Acetic fermentation: Vinegar.

Fibers

food composition—carbohydrates

What are they? A set of complex carbohydrates (polysaccharides) that are barely digestible. They have the properties of hydrocolloids.

They can be divided into

- **Insoluble in water:** These are found in vegetables. The most important insoluble fibers are cellulose and lignin.
- **Soluble in water:** Their most important properties are as thickening and gelling agents. They are only partially absorbed. The most important soluble fibers are:
    - **Pectins:** They form the wall of vegetable cells. They are obtained from apples or lemon peel. Marmalades, jams, and jellies would not be possible without pectins, the indispensable help of sugar, and a little acid (for example, lemon juice). Different types of pectins are used as additives (E-440), according to the desired effect—gelling, thickening, or stabilizing.
    - **Gums:** They are found in plants, algae, and microbes and have the capacity to increase the viscosity of liquids, and to form gels.

Most Common Fibers		
Insoluble in water	Cellulose	
	Lignin	
Soluble in water	Gums	Carobin
		Gellan
		Tragacanth
		Agar
		Carrageenans
		Xanthan
		Alginate
		Others
	Pectins	HM
		LM

Fibrous proteins

food composition—proteins

What are they? Proteins that have a simple structure of amino acids (polypeptide chains) organized in one direction and normally in parallel arrays.

Additional information:

- They are abundant in superior animals such as vertebrates.
- Two important classes of fibrous proteins are considered here: keratins (proteins in the skin, wool, nails, etc.) and collagens (proteins in the tendons, muscles, etc.).

Ficin

See **Enzymes**

Filtration

physical or chemical processes

What is it? A procedure to separate a heterogeneous mixture based on the different sizes of its particles, which are filtered through (or not) the orifices of a mesh.

For example: Orange juice can be separated from its pulp by filtration.

Types of filtrations:

- **According to the orientation of the mixture of components:**
  - **Vertical filtration:** where the mixture passes vertically through filters:

Component mixture

↓

Filter

↓

Filtrate

The process can be speeded up by producing a depression (making a vacuum) or by exerting pressure on the mixture of components.

- **Tangential filtration:** where the mixture passes tangentially and under pressure through the filter. This method is used for more selective separations.

  Component mixture
  ↘
  Filter
  ↓
  Filtrate

This filtration process even allows separation in a molecular environment. For example: it allows water to be separated from wine (concentration).

According to the filter used:

- Mesh: Separates particles up to 0.001 mm.
- Microfilter: Separates particles up to 0.000001 mm.
- Nanofilter: Separates particles up to 0.000000001 mm.
- Ultrafilter: Separates particles up to 0.000000000001 mm.

Flavonoids

food composition—pigments and other compounds

What are they? They are polyphenol-type vegetable compounds, responsible for the coloration of many fruits and vegetables.

Where do they come from? How are they obtained? They are extracted from vegetable products.

Form: Liquid or powder.

Additional information:

- The most important flavonoids within the polyphenols are anthocyans and tannins. *See* Anthocyans, Tannins.
- One of the tannin groups can be classified as flavonoids.

General uses:

- **In the retail food industry:** As nutraceutical foodstuffs.
- **In restaurants:** No known use.

Flavor

organoleptic perceptions

What is it? A term that usually refers to taste but which has recently been used to describe the ensemble of organoleptic perceptions (smell, taste, touch) produced by a foodstuff.

The definition given by *Merriam-Webster* (3rd version) is "the blend of taste and smell sensations evoked by a substance in the mouth." This definition does not include tactile sensations.

Flavor enhancer

food concepts

What is it? A product that when added to a foodstuff increases its organoleptic properties, especially taste.

Examples: Salt (sodium chloride), sugar, sodium glutamate, etc.

Foam

physical or chemical processes

What is it?

Scientifically:

- A colloidal dispersion of a gas in a liquid (water G/W or oil G/O), where G represents the gas dispersed in W (water) or O (oil); for example, beer froth (G/W).
- A colloidal dispersion of a gas in a solid (G/S), where G represents the gas dispersed in S, a solid; for example, soufflé, bread, sponges, etc.

Gastronomically:

- According to the nomenclature used to designate the preparations created in the elBulli restaurant, a foam is a preparation with a variable texture, normally very light, obtained from a purée or jellied liquid that is placed in a siphon. Recently, this name has been to other preparations made in the siphon, even though other ingredients may be added (egg white or yolk, cream, fats, starches, etc.). They may be hot or cold.

Foaming agents
food concepts

What are they? Products that allow the formation of foam from a liquid foodstuff.

For example: Quillaia (E-999, extract from the soapbark tree, *Quillaja saponaria*). The glycirrhizins from liquorice and, in general, all saponins and amino acids are foaming agents.

Food composition
food concepts

What is it? The list of ingredients that a foodstuff is composed of, expressed normally in percent (grams per 100 g of food) of its content. The normal sections in the food composition tables are as follows:

Examples of Foodstuffs with a Summary of Their Composition			
	Almonds	Veal	Oranges
Water	5%	73.4%	88.2%
Proteins	19%	19%	0.9%
Lipids	53.1%	5.9%	0%
Carbohydrates	19.5%	0%	8.6%
Mineral salts, vitamins, and others	3.4%	1.7%	2.3%

Food esters
food concepts

What are they? Substances formed by the reaction of an organic acid with an alcohol that constitute an important part of the aromas of foodstuffs.

For example:

- Isoamyl acetate is mainly responsible for the aroma of bananas.
- Ethyl butyrate is mainly responsible for the aroma of pineapples.
- Octyl acetate is mainly responsible for the aroma of oranges.

Foodstuffs
food concepts

What are they? All substances or products, whether solid or liquid, natural or processed, that may be used for human nutrition due to their characteristics, applications, composition, preparation and state of preservation.

Additional information: They provide living beings with energy and raw materials that form body tissues and organs as well as chemical compounds, such as minerals and vitamins, to regulate vital functions.

Freezing
physical or chemical processes

What is it? A physical process during which a product is submitted to a temperature that is low enough to turn the liquid it contains, normally water or a aqueous solution, into a solid.

Additional information:

- In freezing conditions the decomposition reactions of the food slow down so much that the lifespan of the frozen product is prolonged.
- If freezing takes place at very low temperatures, it is called ultra freeze (temperature lower than –180°C). *See* Fusion point

Fructose
food composition—carbohydrates

What is it? A simple carbohydrate (monsaccharide) that functions as a sweetener.

Where does it come from? How is it obtained? It is extracted from fruit, honey, etc.

Form: Powder.

Additional information:

- Also called levulose.
- Sweetening power: Approximately 1.1–1.7 times the sweetness of sugar (sucrose).
- It can be found in abundance in honey. It also forms part of many fruits such as apples (60% of the sugars), figs (40% of the sugars) and grapes (40% of the sugars). It can also be found in other vegetable

products such as: tomatoes (60% of the sugars), cabbages (30% of the sugars) and carrots (20% of the sugars).

General uses:

- **In the retail food industry:** Confectionery, drinks, chewing gum, jams, etc.
- **In restaurants:** No known direct use. It is used indirectly in some foodstuffs as honey, invert sugar, and in corn syrup.

Fungi

scientific concepts

What is it? One of the kingdoms of living beings. It includes yeast, mold, mushrooms, etc.

Additional information:

- Mushrooms are used for direct consumption (not all are edible) and yeast is used in fermentations.
- The mold produced by foodstuffs, such as cheeses, to show their state of putrefaction, is also considered to be a fungus.

Furcellaran (E-407a)

additives—gelling agents
additives—thickening agents
additives—stabilizers.

What is it? A carrageenan that is used as a gelling, thickening, and stabilizing additive. It has the properties of a hydrocolloid.

Where does it come from? How is it obtained? It is extracted from the algae *Furcellaria fastigiata*.

Form: Powder.

Additional information:

Similar to the kappa carrageenan.

General uses:

- **In the retail food industry:** It is used best in dairy desserts, cream, ice creams, etc.
- **In restaurants:** It is used in dairy desserts.

Fusion point (enthalpy)
scientific concepts

What is it? The temperature at which a substance transforms from solid to liquid or vice versa. It is also referred to as the freezing or solidification point. It depends very little on atmospheric pressure.

Additional information:

- The fusion point is an important property in the culinary treatment of fats, chocolate coatings, gelatin, etc.; for example, the fusion point of chocolate and gelatin (used in a product) is roughly the human body temperature (34–36°C), which is why they melt in the mouth.

Examples:

- The fusion-freezing point of distilled water (with no other dissolved component) is 0°C. If salt or sugar are dissolved in it, the freezing point reduces in accordance with the concentration of the dissolved components. This property is used in antifreezes, where substances that are dissolved in the water enable the product to be cooled below 0°C without forming ice. When temperatures are low, roads are salted under the same principle.
- Ethyl alcohol has a fusion point of –114.6°C (at atmospheric pressure). This is why it is practically impossible to freeze alcoholic beverages, especially spirits. Obviously, it is possible to do so with liquid nitrogen.
- Mixtures of substances with similar chemical compositions (for example, oils and fats) do not have a defined fusion point, rather an interval of fusion points.

Product	Fusion Point	Boiling Point
Nitrogen	–209°C	–195.8°C
Ethyl alcohol	–114.6°C	78.5°C
Carbon dioxide	–78°C	Sublime
Water	0°C	100°C
Salt (sodium chloride)	800°C	1,442°C

Fusion temperature

See **Fusion point.**

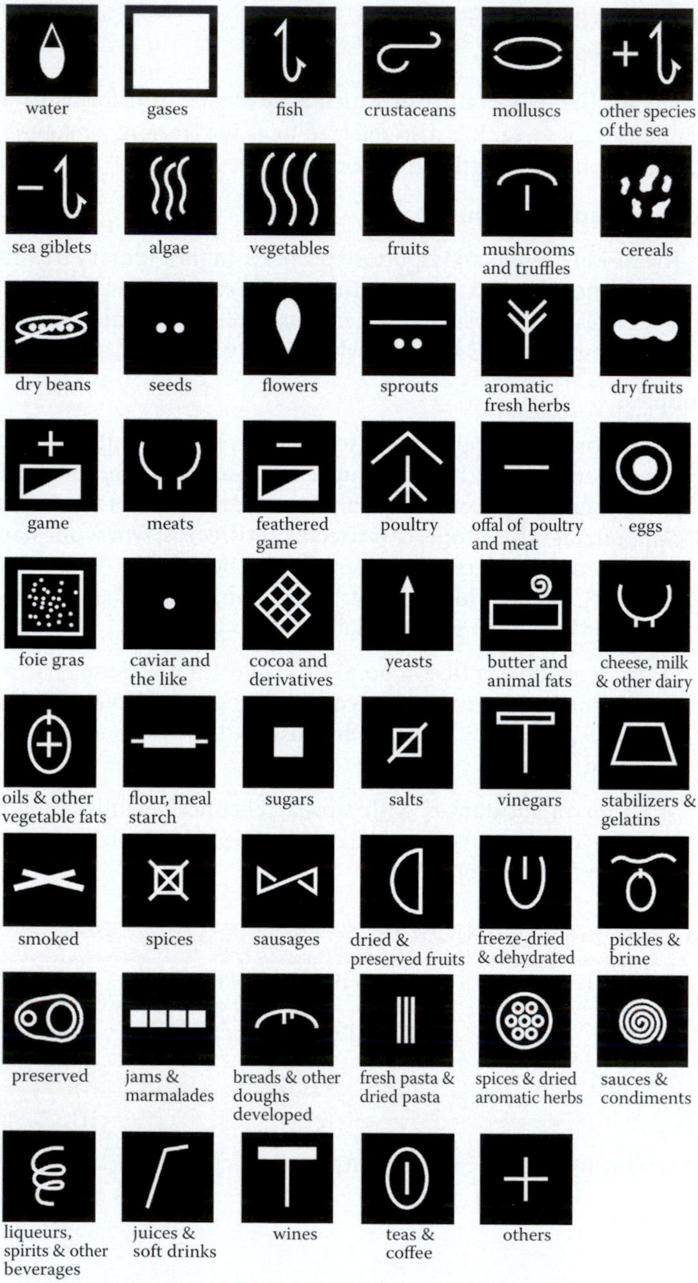
water
gases
fish
crustaceans
molluscs
other species of the sea
sea giblets
algae
vegetables
fruits
mushrooms and truffles
cereals
dry beans
seeds
flowers
sprouts
aromatic fresh herbs
dry fruits
game
meats
feathered game
poultry
offal of poultry and meat
eggs
foie gras
caviar and the like
cocoa and derivatives
yeasts
butter and animal fats
cheese, milk & other dairy
oils & other vegetable fats
flour, meal starch
sugars
salts
vinegars
stabilizers & gelatins
smoked
spices
sausages
dried & preserved fruits
freeze-dried & dehydrated
pickles & brine
preserved
jams & marmalades
breads & other doughs developed
fresh pasta & dried pasta
spices & dried aromatic herbs
sauces & condiments
liqueurs, spirits & other beverages
juices & soft drinks
wines
teas & coffee
others

Pictograms that represent the food families used in cooking. This systematization was created in 2001 and was inspired by the writing of the book *elBulli 1998–2002*.

G

Galactomannans

additives—thickening agents

What are they? Fibrous carbohydrates that are used as thickening additives. They have the properties of hydrocolloids.

Additional information:

- Chemically, they are a group of gums formed by galactose and mannose (hence their name). The most important soluble fibers are carob, guar, and tara. *See* Carob gum, Guar gum, and Tara gum.

Galactose

food composition—carbohydrates

What is it? A simple carbohydrate (monosaccharide) that forms lactose when combined with glucose in a certain way.

Additional information:

- It can also be found in other carbohydrates, such as carob gum.
- Its sweetening power is 0.3–0.5 times that of sucrose (sugar).

Gases (additives)

additives—gases

What are they? Food additives that are found in a gaseous state at normal temperature for food consumption.

Types:

- Argon (E-938).
- Helium (E-939).

- Nitrogen (E-941).
- Nitrous oxide (protoxide of nitrogen, dinitrogen monoxide E-942).
- Oxygen (E-948).
- Carbon dioxide (E-290).

General uses:

- **In the retail food industry:**
    - **Packaging:** Placed with foodstuffs in packaging for basic preservation functions. The technique is usually referred to as MAP (modified atmosphere packaging); for example, nitrogen gas and sometimes carbon dioxide.
    - **Stability:** They prevent foodstuffs from losing their characteristics (replacing "air" because of oxidation concerns); for example, nitrogen gas in sauces.
    - **Shape:** They prevent foodstuff containers from losing their shape. For example: a little gas (normally nitrogen) is placed in tetra brik containers so that they keep their shape (it prevents the walls from being sucked in by the vacuum that may be formed).
- **In restaurants:**
    - The nitrogen protoxide (N_2O) is used in siphons to give a foam consistency to the product within.
    - In the case of carbon dioxide (CO_2), gas is placed in soda siphons.

Gasifying agents

food concepts

What are they? Products or mixes of products that release gas and consequentially increase the volume of foodstuffs.

Additional information:

- The mixture of bicarbonate (bicarbonate of soda) and an acid (for example, citric acid) also produces carbon dioxide. It is the combination used for effervescent tablets and other gaseous effects. These mixtures are called baking powders or chemical yeasts. *See* Carbon dioxide.

Gel

scientific concepts

What is it? A colloidal dispersion of a liquid in a solid. It is characterized by the creation of a tridimensional network structure; for example, crème caramel, jellies formed by gelling agents, etc.

Gelatin (gelatine)

food composition—proteins

What is it? A mixture formed by proteins that are soluble in water that is traditionally used as a gelling agent. It has the properties of a hydrocolloid.

Where does it come from? How is it obtained? By physicochemical separation of bone or skin collagen, mainly from pork but also from veal.

Form: Leaves or powder.

Additional information:

- In addition to being a gelling agent, it is also an emulsifying agent.
- Many types of gelatin are measured in bloom values. The higher this figure, the harder and more consistent the gel will be. The most widely used is bloom 220.
- Gelatin exists in the form of leaves (2 g per unit), which should be hydrated in cold water and heated in the product to 40–50°C. It can also be found in powder form.
- If gelatin has received a pregelatinization treatment, it does not need to be heated at a temperature to achieve the gelling effect.
- Until recently, it was the most commonly used gelling agent in western cooking.

General uses:

- **In the retail food industry:** Dairy-product desserts, pastries, meats, product restructuring, etc.
- **In restaurants:** Many applications, the most important in cold jellies, both sweet and savory. Also used as an emulsifying agent to produce foams.

Maximum/minimum quantity: QS (minimum quantity required to obtain desired effect).

Gellan Gum (E-418)

additives—gelling agents
additives—stabilizers

What is it? A fibrous carbohydrate that is used as a gelling additive and a stabilizer. It has the properties of a hydrocolloid.

Where does it come from? How is it obtained? Fermentation of carbohydrates by the bacterium *Sphingomonas elodea*.

Form: Powder.

Additional information:

- Discovered in 1977.
- Two types can be distinguished: rigid and elastic.
- It forms gels in the presence of calcium or very low concentrations of acids. If the foodstuffs to be jellied contain calcium, the effect is better and more consistent.
- It resists high temperatures, as does agar-agar.
- Elastic gellan gum has been used to obtain suspension effects so that objects can be suspended in liquid.

General uses:

- **In the retail food industry:** Ice creams, jams, and drinks with suspending powers. It is little used in the retail food industry because of its high price.
- **In restaurants:** Ice creams and other dairy desserts, with some use in jams and jellies.

Gelling agents

food concepts

What are they? Products that form gels that give texture to foodstuffs. Its properties define it as belonging to the hydrocolloid family.

Most Common Gelling Agents

Carbohydrates (fibers)	Plants (cellulose)	Methylcellulose	
	Plants (resin exuding)	Karaya	
	Plants (tubers)	Konjac	
	Microbes	Curdlan	
		Gellan	Elastic
			Rigid
	Plants (fruits)	Pectins	LM and LA
			HM
	Algae	Carrageenans	Iota
			Furcellaran
			Kappa
		Alginates	
		Agar-agar	
Proteins	Vegetables		
	Animals	Gelatin (leaves or powder)	
		Others	

Gene

scientific concepts

What is it? Each and every part of the chromosome of a cell that contains information regarding a characteristic or part of a characteristic of a living being.

Additional information:

- Genes that control smell are capable of defining some 10,000 different variations, but not everybody can detect them all, nor with the same intensity.
- The product phenylthiourea has an extremely bitter taste for part of the population, whereas it is insipid for others. This is determined by a gene.

Genetics
scientific concepts

What is it? A science that studies hereditary information and how it is transmitted. Regarding food, all people have different organoleptic connotations, which is shown in the tasting of different products. Extreme cases are genetic intolerance diseases; for example, gluten, lactose intolerance, etc.

Genome
scientific concepts

What is it? A set of genes that form the genetic code; in other words, the information required to constitute a living being and coordinate its development.

Additional information:

- The genome is made up of between 25,000 and 30,000 genes in the human species.
- All foodstuffs have their own genome, and knowledge of this will enable the understanding of characteristics related to their organoleptic properties.

Globular proteins
food composition—proteins

What are they? They are more complex than fibrous proteins as the amino acids are linked into chain structures that are also linked together in some cases.

Additional information:

- Hemoglobin in blood is a complex protein (globular).
- The main globular proteins are
  - **Nutritional structural proteins:** There are many proteins that we can associate with this group, from the myoglobin of the muscles to those that form wheat flour.
  - **Myoglobin:** A relatively small globular protein that is found in muscles and is especially abundant in marine mammals such as the whale.

- **Wheat flour proteins:** Wheat flour has two types of basic components, starch and proteins.
- **Enzymes:** *See* Enzymes.

Glucose
food composition—carbohydrates

What is it?

- **Scientifically:** A sugar-type simple carbohydrate (monsaccharide) used as sweetener. Referred to in gastronomy and the retail food industry as dextrose.
- **Gastronomically:** It is the name given to glucose syrup.

Where does it come from? How is it obtained? It is extracted from fruit, honey, etc. and industrially from starch.

Form: Powder.

Additional information:

- Its sweetening power is 0.5–0.8 times that of sucrose (sugar).
- It should not be confused with the glucose used in the kitchen, i.e., glucose syrup.

General uses:

- **In the retail food industry:** Confectionery, drinks, chewing gum, jams, a sweetener for use by diabetics.
- **In restaurants:** In sweet preparations.

Quantity and instructions for use:

- **Basic quantity for cooking:** QS.
- **Instructions for use:** Mix directly with the ingredients.

Glucose syrup
food composition—carbohydrates

What is it? A mixture of carbohydrates formed by glucose (dextrose), maltose, triose, and other superior carbohydrates, and used as a sweetener.

Where does it come from? How is it obtained? From the breakdown of starch chains.

Form: Thick liquid or powder (atomized glucose).

Additional information:

- Sweetening power: 0.3–0.5 times that of sugar (sucrose).
- Glucose syrups should have, according to the EU, a minimum of 20 DE (dextrose equivalent).
- Initially, glucose syrups were 38 DE, but today, with the use of enzymes that break them down, the following types can be found:
    - Low conversion syrups, between 20 and 38 DE; for example, caramel-type preparations.
    - Regular conversion syrups, between 38 and 55 DE. The most widely used product in cooking is DE 35–40; for example, gummies.
    - High conversion syrups, between 55 and 80 DE (that have a certain proportion of fructose); for example, marshmallows.

General uses:

- **In the retail food industry:** Confectionery, drinks, chewing gum, jams, etc.
- **In restaurants:** In sweet preparations.

Quantity and instructions for use:

- **Maximum/minimum quantity:** QS (minimum quantity required to obtain desired effect).
- **Basic quantity for cooking:** It can constitute nearly 100% of the product.
- **Instructions for use:** Mix the ingredients together directly and heat to the proposed temperature.

Glutamate (E-621)

additives—flavor enhancers

What is it? An additive derived from the amino acid glutamic acid (GLU), used as a flavor enhancer for foodstuffs, especially meat.

Where does it come from? How is it obtained? It is extracted from animal or vegetable proteins and neutralized to transform from glutamic acid to glutamate.

Additional information:

- The most widely used glutamate is sodium glutamate monosodium glutamate or MSG (E-621).

- It is widely used in Chinese and Japanese cooking.
- It is a product associated with what is still a controversial taste (umami), although it is considered to be one of the basic tastes in some Asian cultures.
- Guanylate and inosinate are similar products used in the food aroma industry.
- Some products contain glutamate in their composition: tomato, some algae, etc.

General uses:

- **In the retail food industry:** To increase the flavor of stocks, meats, etc.
- **In restaurants:** It is rarely used in western cooking, but ingredients naturally containing glutamine are used frequently.

Gluten

food composition—proteins

What is it? A set of proteins that forms part of the flours of some cereals.

Where does it come from? How is it obtained? Separation of non-proteinaceous components from wheat flour (but also from oats, barley, etc.).

Form: Powder or granulated.

Additional information:

- It is the principal protein of wheat.
- Strong flours have the highest percentage of gluten.
- The gluten intolerance suffered by some people is called celiac disease.
- It is formed by a mixture of proteins, mainly gliadins and glutenins. These proteins are responsible for the formation of viscoelastic dough capable of capturing gas during fermentation. Therefore, the quantity of gluten in flour is very important in breadmaking, because if the flour is lacking in gluten, the bread will not be springy.
- Its elastic properties and its ability to spread out enable doughs to be mixed in the breadmaking process, and contribute to the retention of carbon dioxide in the fermentation, which makes the dough springy.
- Flours have a protein proportion of 8–15% (of which 85% is gluten). In strong flours, it is somewhat higher (11–15% proteins) than weak flours (8–11%).

- With regard to flour, its properties produce:
    - Flour enrichment (increasing the strength of the flour, increasing water absorption and increasing proteins).
    - Bread improvers.
    - Special breads:
        - Sliced bread: Improves the crumb, the cut, and springiness.
        - Rye bread: Improves volume.
    - Flat and extruded doughs: Improves stretchability.

General uses:

- **In the retail food industry:** It enriches flours, improves bread, special breads, flat and extruded doughs (cannelloni, macaroni, etc.). To increase the protein content of certain compounds, etc.
- **In restaurants:** In breadmaking and dough toughening.

Glycans

food composition—carbohydrates

What are they? A group of complex carbohydrate products (polysaccharides), formed exclusively by glucose chains.

Additional information:

In this group the following should be noted:

- Amylose and amylopectin (components of starch)
- Glycogen
- Cellulose

See Amylopectin, Amylose, Cellulose, Glycogen

Glycerides (E-471, E-472, E-474)

food composition—lipids
additives—emulsifiers

What are they? Compounds formed by the union between fatty acids and glycerin. The blends of glycerides are habitually known as fats and oils. As additives, they are used for their emulsifying properties (monoglycerides E-471, diglycerides E-472 and sucroglycerides E-474).

Types of glycerides:

- **Monoglycerides:** One single molecule of fatty acid per molecule of glycerin.
- **Diglycerides:** Two molecules of fatty acids per molecule of glycerin.
- **Triglycerides:** Three molecules of fatty acids per molecule of glycerin.
- **Others:** For example, sucroglycerides, which are obtained by reacting sucrose with an edible fat or oil.

Additional information:

- Triglycerides are by far the most abundant in nature.

General uses:

- **In the retail food industry:** Widely used in crèmes (emulsifiers and stabilizers), mayonnaises (emulsifiers and stabilizers), chocolates (fluidifiers), breads (they increase the duration of springiness), cream (stabilizers), etc.
- **In restaurants:** In experimentation.

Glycerin (E-422)

food composition—alcohols
additives—stabilizers
additives—humectants

What is it? A structure component of many lipids, also called glycerol and used as a stabilizing and moisturizing additive.

Additional information:

- Unions of glycerin with fatty acids produce glycerides (in lipids). The blends of glycerides are otherwise known as fats and are important, especially triglycerides. Animals and many vegetables synthesize them naturally.
- It is also a component of other lipids, such as lecithin (phospholipid).

General uses:

- **In the retail food industry:** As a humectant in different products.
- **In restaurants:** As a sweetener, humectant, and beverage base.

Glycide

See **Carbohydrates**

Glycirrhizin

food composition—carbohydrates

What is it? A product that belongs to the saponin group, formed basically by chains of carbohydrates.

Where does it come from? How is it obtained? Extracted from vegetable products.

Form: Powder.

Additional information:

- It provides the basic organoleptic properties of licorice.
- It has some properties of foaming agents.
- In low proportions, as in licorice, it is not harmful to the organism, but it is toxic as an individual product.

General uses:

- **In the retail food industry:** It is used as a sweetening agent and aromatizer in pharmaceuticals and as a foaming agent in non-alcoholic drinks.
- **In restaurants:** Used indirectly in the form of licorice to obtain airy preparations.

Glycogen

food composition—carbohydrates

What is it? A digestible complex carbohydrate (polysaccharide), from the glycan family.

Additional information:

- It is a molecule formed by millions (a "package") of molecules of glucose that the animal organism uses as an energy reserve.
- Sugars consumed with foodstuffs are converted into glycogen with the help of insulin. They are accumulated in the liver and in the muscles, and when free glucose is needed, an enzyme is responsible for decomposing the glycogen into glucose ("unpacking").

- It is the energy reserve product of animals. Its structure is similar to starch amylopectin; although more ramified, it is perfectly digestible and we eat it in animal meat, especially liver.

Gold (E-175)

additives—coloring agents

What is it? A mineral product that is used as a coloring agent for coatings.

Where does it come from? How is it obtained? From physical treatment of gold processing.

Form: Powder or sheets.

General uses:

- **In the retail food industry:** To coat and decorate confectionery and in baking.
- **In restaurants:** As a coloring agent for coating products and preparations, although its use is very limited.

Quantity and instructions for use:

- **Basic quantity for cooking:** QS (minimum quantity required to obtain desired effect) for the coating of preparations.
- **Instructions for use:** Brush the surface of the preparation with the coating. Mixing with water or alcohol speeds up the process.

Guar gum (E-412)

additives—thickening agents
additives—stabilizers

What is it? A fibrous carbohydrate from the galactomannans group. An additive used as a thickening agent and stabilizer. It has the properties of a hydrocolloid.

Where does it come from? How is it obtained? From the seeds of a leguminous plant (*Cyamopsis tetragonolobus*) that is similar to the pea and native to India and Pakistan.

Form: Powder.

Additional information:

- The seed of the plant has been used for human and animal nutrition for centuries.

- The gum contained in the seed is soluble in cold water and is very viscous. It is used in products that must be submitted to high temperatures and, in general, as a complement to other thickening additives.

General uses:

- **In the retail food industry:** Unaged cheeses, ice creams, croquettes, sauces, bakery products, preserves, jellies, jams, etc.
- **In restaurants:** Ice creams, sauces, jellies and jams.

Gum

food concepts

What is it? A group of complex fibrous carbohydrate products (polysaccharides). These carbohydrates have the properties of hydrocolloids.

Additional information:

- They are hydrocolloids and therefore act as thickening and gelling agents.
- They are categorized as food additives.

Most Common Gums	
Designated as Gums	**Not Designated as Gums**
Carobin Gum	Agar-agar
Guar Gum	Alginates
Tragacanth Gum	Carrageenans
Arabic Gum	
Xanthan Gum	
Karaya Gum	
Tara Gum	
Gellan Gum	
Konjac Gum	

See Fibers.

H

Hardness (of water)
scientific concepts

What is it? The content of calcium and magnesium salts in water.

Additional information:

Hard water is present in areas where rocks are calcareous. In hard-water areas, calcium coats vegetables, which then need to be cooked longer and a substance that precipitates the calcium salts needs to be added.

Helium (E-939)
additives—preservatives
additives—gases

What is it? An inert gas of the argon family, used as a preservative and gasifying agent.

Where does it come from? How is it obtained? From deposits. It is extracted or separated from gas.

Form: It is compressed in cylinders of varying sizes, usually 25 liters.

Additional information:

- It is present in some natural gases.
- It is used as a gas propellant (in balloons).
- Its use in restaurants is complicated because it tends to escape from the place where it is supposed to be contained. A high-pitched sound is produced when it comes into contact with vocal chords, without being harmful to health.

General uses:

- **In the retail food industry:** Modified atmospheres packaging.
- **In restaurants:** No known use.

Hemoglobin
food composition—proteins

What is it? A red-colored protein, with the function of transporting oxygen through the blood stream in the organism.

Additional information:

- The heme group contains iron and is surrounded by an amino acid structure that forms the globins, which are used to transport the heme.
- This protein is of interest to the retail food industry and gastronomy because of its emulsifying and gelling effects; for example, blood proteins in the preparation of sausages.

Hexose
food composition—carbohydrates

What is it? A simple carbohydrate (monsaccharide) that has 6 carbon atoms per molecule.

Additional information:

Most simple carbohydrates (monosaccharides) are hexoses; for example, glucose (dextrose), galactose, fructose (levulose), mannose, etc.

HLB (hydrophile/lipophile balance)
scientific concepts

What is it? A number that indicates whether an emulsifier should be used in an oil base (between 0 and 10) or in a water base (between 10 and 20).

Additional information:

- For example, an HLB 3 monoglyceride emulsifier should be dissolved in oil or fat and used. If attempts are made to dissolve it directly in water, an emulsion will not form. However, an HLB 16 emulsifying sucrose ester can be dissolved in water and used.

- The typical scale from 0 to 20, which shows the function and, therefore, use of emulsifiers in specific cases:

HLB	Product Type
From 0 to 2	Anti-foaming agents
From 3 to 6	Emulsifiers that generate an external phase oil emulsion (W/O)
From 7 to 9	Moisturizing agents
From 8 to 18	Emulsifiers that generate an external phase water emulsion (O/W)
From 13 to 15	Detergents
From 16 to 20	Solubilizers of oils in water

HM pectin (E-440)

food composition—carbohydrates
additives—gelling agents

What is it? A soluble, fibrous-type complex carbohydrate (polysaccharide) that is used as a gelling additive. It has the properties of a hydrocolloid.

Where does it come from? How is it obtained? It is present in the cell wall of vegetables

Form: Powder.

Additional information:

- Its name comes from the Greek word *pektos,* meaning strong, solid, firm.
- It is obtained from apples and the peel of lemons or other citrus fruit. The production of jams and marmalades, quince jelly, etc., would not be possible without pectins, the indispensable help of sugar and, sometimes, a little acid (for example, lemon juice).
- An acidic medium is required for this application (pH less than 3.8) and a high proportion of sugars (minimum 60 Brix).

- It gives a thermoirreversible gel that can be heated in the oven without being destroyed (as opposed to LM pectin).

General uses:

- **In the retail food industry:** Jams, preserves, sugar-based pastry products, confectionery, and dairy products.
- **In restaurants:** Fruit pastes, jellies, jams, and baked goods in general.

Quantity and instructions for use:

- **Maximum/minimum quantity:** Fruit juices and nectars: 3 g/L. QS (minimum quantity required to obtain desired effect) in other applications.
- **Basic quantity for cooking:** From 1%; in other words, 1 g per 100 g of liquid to be jellified (10 g per kg).
- **Instructions for use:** It is blended by stirring and heated to boiling point. An acidic medium should be added. Gelling begins between 50 and 60°C but it acquires a gel consistency very quickly. Once jellied, it can be served hot, as it is an irreversible gel. If a more consistent gel is required, the dosage should be increased.

Homogenization

physical or chemical processes

What is it? A procedure to make uniform a mixture of substances that are in suspension or in a colloidal emulsion.

For example: Milk is homogenized by passing the product through orifices with small diameters to break down and reduce the size of fat globules.

Homogenizer

technology—devices

What is it? A device in which a colloidal suspension may be homogenized.

Additional information:

- Very useful for suspensions of hydrocolloids in water.
- Different kinds of disks are rotated, depending on the products to be homogenized.

Hydration

physical or chemical processes

What is it? An increase in the water content of a product.

For example: Hydration of dried wild mushrooms (rehydration). Some products are hydrated when cooked in liquid (rice, pasta, beans, peas, legumes, etc.)

Additional information:

When hydrocolloids absorb water they gain gelling or thickening properties; for example, hydration of gelatin with cold water.

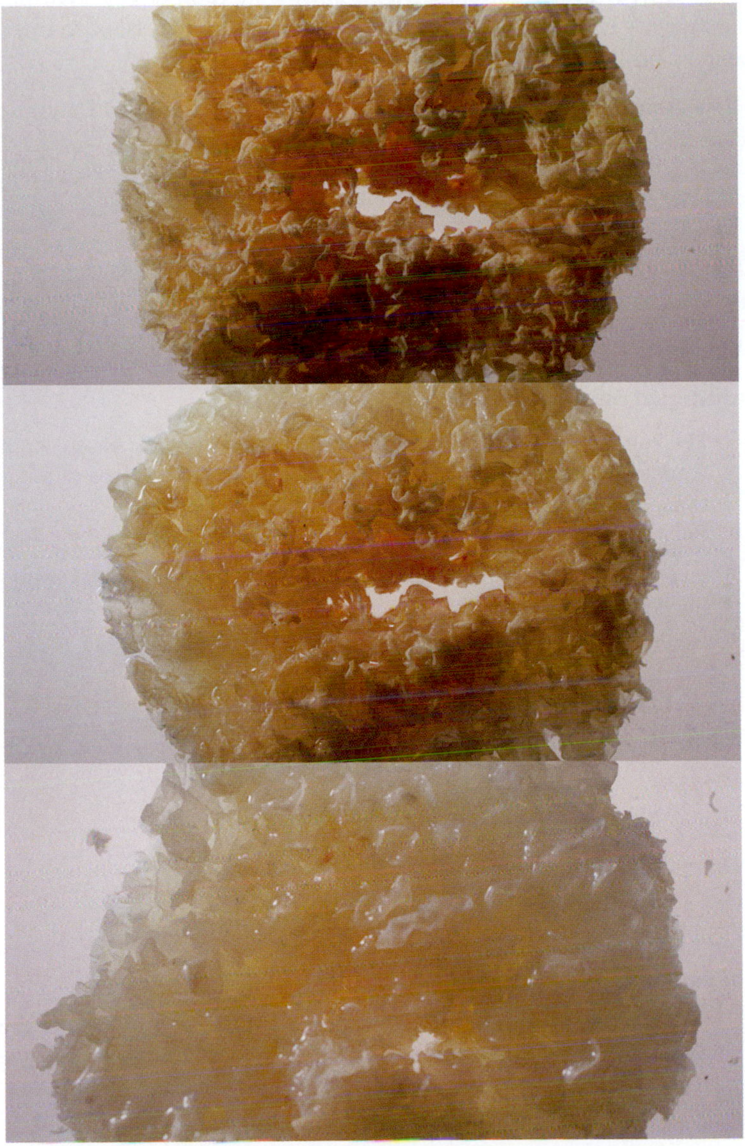

Hydration of the wild mushroom *Tremella fuciformis*.

Hydrocolloid
food concepts

What is it? A protein or complex carbohydrate (polysaccharide) that has the capacity to attract water, causing the formation of gels, or to thicken a blended product or a liquid.

For example: Gelatin leaves (proteins), agar-agar (carbohydrate), pectin (carbohydrate), etc.

Hydrogenation
physical or chemical processes

What is it? A reaction that takes place by the addition of hydrogen.

Additional information:

- Hydrogenation is the industrial process to harden fats, based on the transformation of unsaturated fatty acids into saturated acids by the addition of hydrogen with a catalyst.
- Vegetable oils that are hydrogenated to become margarines are an example. Normally between 15 and 25% of the possible total is hydrogenated.
- Carbohydrates are also hydrated to obtain polyols.

Hydrolysis
physical or chemical processes

What is it? A process whereby a complex substance is chemically broken down by the action of water. This effect is encouraged by acids and some enzymes.

For example:

- When water is added, fats can release fatty acid compounds that cause an unpleasant taste and smell (rancidity). This occurs over time or by certain enzyme actions that can accelerate the process.
- Sugar hydrolysis (sucrose) when an acid (for example, lemon juice) is added. Sugar (sucrose) is broken down to give glucose (dextrose) and fructose. The product obtained is a mixture of glucose and fructose, and is called invert sugar. In the retail food industry, it is used in the caramelization of sugar for crème caramel, etc. and it

provides better control when obtaining dextrose and fructose than caramelization at a higher temperature.

Hydrophile
scientific concepts

What is it? A product that tends to bind water and is soluble in this component; for example, salt (sodium chloride), sugar (sucrose).

Additional information:

- A molecule or part of a molecule that has an affinity for water is also called a hydrophile; for example, one part of lecithin is soluble in water.
- A hydrophile product is lipophobe, i.e., it has no affinity with fats.

Hydrophobe
scientific concepts

What is it? A product that has no affinity with water, and is therefore not soluble in it; for example, fats, oils, etc.

Additional information:

- A molecule or part of a molecule that has no affinity for water is also called a hydrophobe; for example, one part of lecithin is not soluble in water.
- A hydrophobe product is lipophile, i.e., it has an affinity for fats.

Hygroscopicity
scientific concepts

What is it? The property of some products to retain and absorb humidity.

Additional information:

- Sugar-type carbohydrates are hygroscopic. This can create problems when they are used in a very humid atmosphere (lumps form).
- The following is a list of sugars, in ascending order of hygroscopicity:
  - Sucrose (not very hygroscopic)

- Glucose syrups (moderately hygroscopic)
- Dextrose or glucose (fairly hygroscopic)
- Fructose (highly hygroscopic)

There are other hygroscopic products that are not carbohydrates (for example, salt).

I

Identical natural products
scientific concepts

What are they? Synthetic products that have a chemical structure equal to their natural counterparts.

Additional information:

- These substances are produced because the extraction of the natural product has a high financial as well as technical cost.
- For example: Vitamin A (retinol), found in some foodstuffs such as carrots, can be artificially obtained by synthesis, giving us the same product as that produced by nature.

Ingredients (food)
food concepts

What are they?

- **Scientifically:** Any single product that is used to make up a foodstuff.
- **Gastronomically:** A product used to prepare a dish.

Inorganic
See **Chemical compound**

Intensive sweetener

food concepts

What is it? A name given to a series of additives that are used as high-intensity sweeteners.

Additional information:

- Non-energy sweeteners that are added in small quantities to foodstuffs.
- Very intense sweetness. Much greater than sugar (sucrose).
- They are sugar substitutes in foodstuffs for diabetics or low-calorie diets.
- They are not hygroscopic (they do not absorb water).
- They do not add texture to products.
- They are all catalogued as additives.

Intermediate food products (IFPs)

food concepts

What are they? A concept used in the retail food industry to describe products that are rarely ever consumed alone, usually only through the foodstuffs in which they are contained.

Additional information:

- They are chemical compounds or, occasionally, a mixture of a few chemical compounds.
- They can be classified into groups:
  - Complementary (fractions of natural products), for example, lactose.
  - Nutraceutical products or "healthy fractions," for example, flavonoids. *See* Nutraceutical products.

Intestinal flora

scientific concepts

What is it? A set of bacteria in the large intestine that contributes to the absorption of foodstuffs and are therefore, as a whole, beneficial to the organism.

Additional information:

These microbes are capable of partly breaking down the structures of foodstuffs by transforming them into smaller molecules that reach the large intestine where they can then be absorbed by the organism.

Inverse osmosis

physical or chemical processes

What is it? A phenomenon whereby water migrates through a semipermeable membrane. However, unlike osmosis, water travels from the substance with greater concentration to the substance with lesser concentration.

Additional information:

- This unnatural process takes place when strong pressure is exerted on the least concentrated solution.
- Examples of use: Desalination of seawater, elimination of lime in water for domestic use, and in fruit juice concentrations, etc.

Invert sugar

food composition—carbohydrates

What is it? Sugar (sucrose) that has been reduced into glucose (dextrose) and fructose (levulose) by acidic or microbial actions.

Form: Thick liquid.

Additional information:

- This separation into glucose and fructose occurs because of the actions of acids or enzyme invertase derived from yeasts. The result is a thick syrup that contains equal parts of glucose and fructose but with an increased sweetening power (1.25 times sweeter than the initial sugar).
- Honey is composed mainly of invert sugar, which is why it has a thick liquid state. Other compounds give it color and taste.
- Invert sugar would therefore be a type of honey half composed of glucose (dextrose) and half of fructose (levulose). It is used in baking to achieve products that are more liquid and to avoid circumstances where the sugar hardens or forms crystals over time.

- In commercially sold invert sugar, a proportion of sucrose is not inverted.
- It is also referred to as hydrogenated glucose syrup.

Iodine

food composition—minerals

What is it? A chemical element that is a component of certain mineral salts (iodides) and used as a food complement.

Where does it come from? From food products: sea species, especially algae.

Additional information:

- In the human body, iodine is absorbed by the digestive tract and directed to the thyroid glands.
- A lack of iodine is related to brain damage in fetuses and babies, and to psychomotor problems in children.

Iodized

organoleptic perceptions
food concepts

What is it? A flavor profile strongly related to seafood (fish, crustaceans, molluscs, algae, etc.).

Additional information:

A product that has been enriched with iodine to prevent goiter, a disease related to the thyroid that affects people in areas where there is little iodine; for example, iodized salt.

Ion

scientific concepts

What is it? An atom or group of atoms with positive or negative electric charge.

Additional information:

- Positive ions are also called cations, for example, the sodium and calcium ions.
- Negative ions are called anions, for example, the chloride, nitrate, and sulphite ions, etc.

Ionization

physical or chemical processes

What is it? A preservation method during which products are submitted to high-energy rays that eliminate possible microorganisms or any other living element that could cause the decomposition of foodstuffs.

Additional information:

- It is a widely used method for bottled water and for certain foodstuffs without additives.
- The process barely modifies organoleptic properties, but can cause changes in some molecules such as vitamin A.

Iota (E-407)

additives—thickening agents
additives—stabilizers
additives—gelling agents

What is it? A fibrous carbohydrate that is used as a thickening, stabilizing and gelling additive. It has the properties of a hydrocolloid.

Where does it come from? How is it obtained? It is extracted from red algae. *See* Carrageenans.

Form: Powder.

Additional information:

- It forms soft, cohesive (not easily disintegrated) and elastic (stretchable) gels. It is thermoreversible (gel–non-gel, depending on the temperature).
- It can be used to suspend solids in jellied masses and keep them in place.
- Moreover, it is practically the only thixotropic gel—if it is destroyed it rebuilds itself over time. This is why it is used in crème caramel

and other jellied products that must be transported and could suffer gel destruction.

General uses:

- **In the retail food industry:** Dairy products, creams, ice creams, and above all, crème caramel. Chewing gums and breath freshener candies.
- **In restaurants:** In ice cream.

Irish moss
food concepts

What is it? An alga that is native to Carragheen (Ireland) and known scientifically as *Chondrus crispus*.

Additional information:

- On boiling this product, a gelling effect is produced.
- The substances extracted from these mosses were called carrageens or carrageenans after the village where they were first used. Today the carrageenans (E-407) are still called Irish moss in some places.

Iron
food composition—minerals

What is it? A chemical element, used as a food complement, which is a component of some mineral salts that are present to a greater or lesser degree in most foodstuffs.

Where does it come from? How is it obtained? It is extracted from food products, mineral salts.

Additional information:

- Its presence in major proteins such as hemoglobin (a protein that transports oxygen through the blood) makes it an important element. Its absence causes anemia.
- This iron is often not assimilable (the organism is not able to absorb it). Vitamin C (ascorbic acid) encourages iron assimilation (absorption). For this reason, certain mixtures have long been prepared to aid iron absorption.

Examples of Foodstuffs	Percentage of Iron
Pork liver	22 mg per 100 g
Kidneys	10 mg per 100 g
Oysters	5.5 mg per 100 g
Eggs	2.7 mg per 100 g
Sardines	2.5 mg per 100 g
Dried dates	2.5 mg per 100 g
Pulses	6.7 mg per 100 g

Isomaltitol (or Isomalt) (E-953)

additives—sweeteners
additives—humectants

What is it? An artificial product from the polyol group, used as a sweetening and moisturizing additive.

Where does it come from? How is it obtained? By synthesis (chemical reaction) from sucrose.

Form: Powder or granulated.

Additional information:

- Sweetness: 0.5 times that of sugar (sucrose).
- It has been used in restaurants because of its great stability in relation to environmental humidity, the lack of which creates a considerable problem in other sweet products. Since the mid-1990s it has been almost the only polyol used in catering.
- It is stable at high temperatures (150°C). This property makes it appropriate for sweet products without giving them the typical color of burnt caramel.
- It has low energy and is therefore used in sugar-free sweets.
- It can have a laxative effect in concentrations higher than 60 g/kg.
- Its very low hygroscopic tendency is very important.

General uses:

- **In the retail food industry:** Confectionery, caramels, chewing gum, jams, sweetener for diabetics, texture preservative.
- **In restaurants:** Caramels and pastry products in general.

Quantity and instructions for use:

- **Maximum/minimum quantity:** QS (minimum quantity required to obtain desired effect).
- **Basic quantity for cooking:** For sugar substitution, use the necessary quantity.
- **Instructions for use:** It is mixed with other sugars when necessary, and is brought to the temperature required by the preparation.

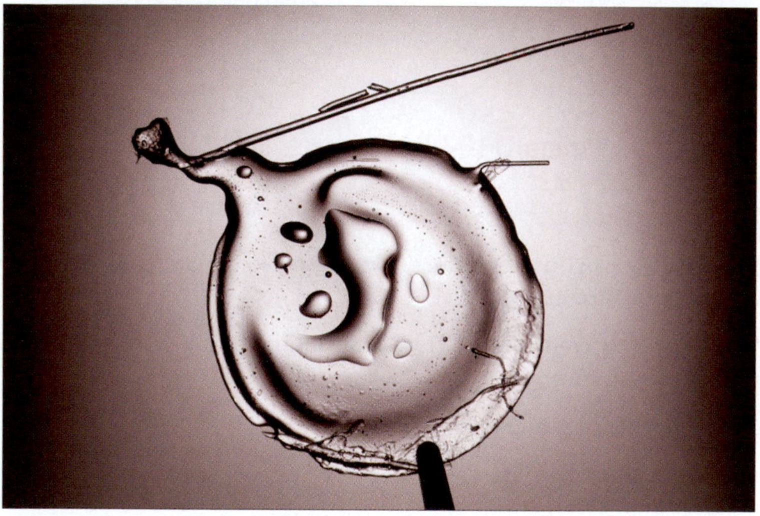

Caramel prepared with isomaltitol.

K

Kaolin

mineral products
additives—anti-caking agent (E-559)

What is it? Hydrated aluminum silicate.

Where does it come from? How is it obtained? Extraction and purification of mineral products.

Form: Powder.

General uses:

- **In the retail food industry:** As an anti-caking agent for powder products.
- **In restaurants:** Used in lipid powders.

Kappa (E-407)

additives—thickening agents
additives—stabilizers
additives—gelling agents

What is it? A fibrous carbohydrate that is used as a thickening, stabilizing and gelling additive. It has the properties of a hydrocolloid.

Where does it come from? How is it obtained? It is extracted from red algae. *See* Carrageenans (or Carrageens) (E-407).

Form: Powder.

Additional information:

- It forms hard, strong, and brittle gels. It is thermoreversible (gel-non-gel, depending on the temperature).
- One of its most important characteristics in cooking is the speed at which the gel can be formed when the temperature is lowered; this facilitates the jellied coating of other products.
- It bonds with casein (milk protein) and is therefore also used to stabilize cocoa suspensions in milk so that the cocoa powder does not remain at the bottom of the container.
- It suffers a little syneresis (water loss) once the gel has been formed. Adding carob gum can reduce this as well as make the gel harder.

General uses:

- **In the retail food industry:** Coatings, chocolate milk shake suspensions, low-calorie preserves, milk desserts, creams, and ice creams. It retains water in meats, fish, and cooked products.
- **In restaurants:** Coatings and cooked meat emulsions.

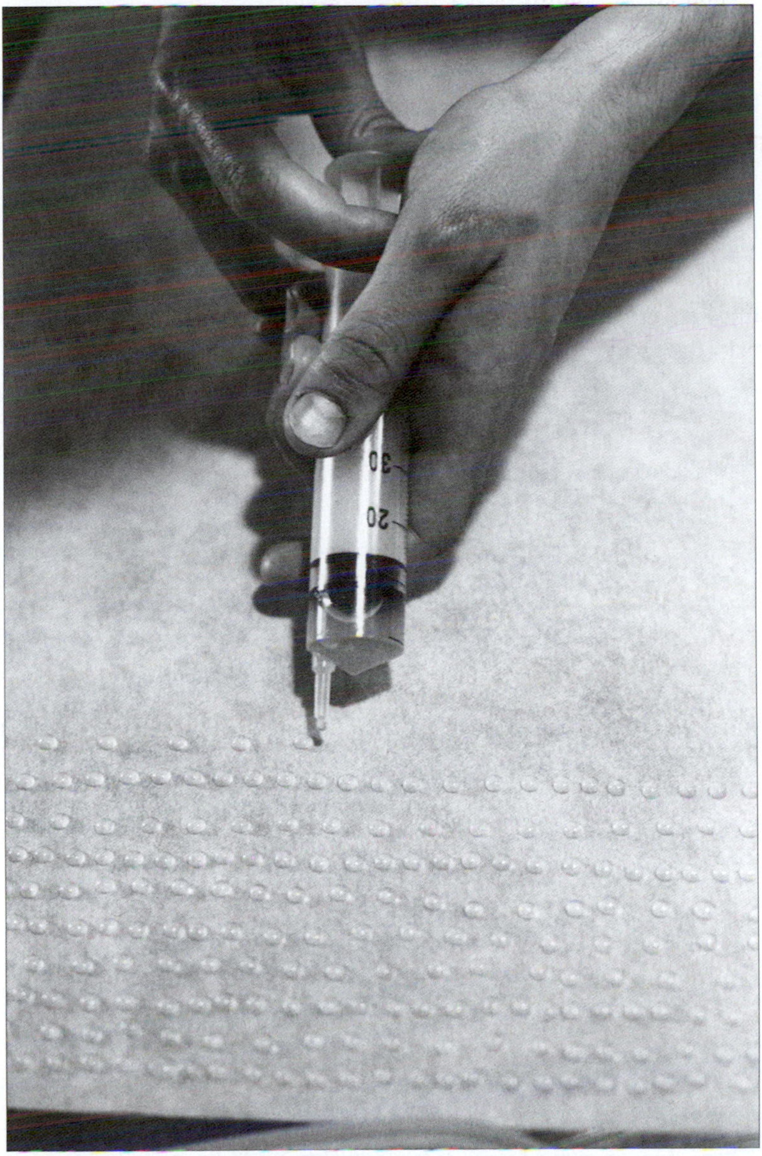
30
20

Citric pearls, made possible by kappa.

Karaya gum (E-416)

additives—thickening agents
additives—stabilizers
additives—gelling agents

What is it? A fibrous carbohydrate that is used as a thickening, stabilizing, and gelling additive. It has the properties of a hydrocolloid.

Where does it come from? How is it obtained? It is extracted from the tree *Sterculia urens* (India, China).

Form: Powder.

Additional information:

It is not very soluble but creates viscous solutions that are stable at low pH values and resistant to heat treatments and freezing.

General uses:

- **In the retail food industry:** A stabilizer in meringues and airy products, ice creams, sorbets, and water ices.
- **In restaurants:** Stabilizer in foams.

Keratins

food composition—proteins

What are they? Animal proteins that are insoluble in water and derived from skin cells and derivatives: hair, nails, wool, scales, feathers, etc.

- Chicken skin is almost entirely composed of keratin-type proteins, as are fish scales.

Konjac gum or flour (E-425)

additives—gelling agents
additives—thickening agents

What is it? A fibrous carbohydrate that is used as a gelling and thickening additive. It has the properties of a hydrocolloid.

Where does it come from? How is it obtained? By grinding the Asiatic tuber *Amorphophalus konjac* that is used in Asia as a foodstuff.

Form: Powder.

Additional information:

- It is forbidden in children's sweets because of the problems caused by Chinese jellies with excessively high levels of konjac because konjac does not melt in the mouth but must be chewed.
- Much used in Asian cooking, particularly in Japan.

General uses:

- **In the retail food industry:** Pasta for soups, restructured products (meat, etc.), fat substitute, etc.
- **In restaurants:** Used in sweets.

L

Lactic acid (E-270)

additives—preservatives
additives—acidity regulators
food composition—acids

What is it? An organic acid that is present in fermented milk products. It is used as an acidity regulator and as a preservative.

Where does it come from? How is it obtained? From the fermentation of sugars in milk and dairy products. But it also can be obtained by fermentation of carbohydrates using bacteria.

Form: Liquid.

Additional information:

- It is formed in muscle tissue after prolonged physical effort.

Note: The burning sensation in muscles during prolonged physical effort used to be attributed to the production of lactic acid, but muscle soreness and stiffness after strenuous exercise is actually due to small injuries of the muscle fibers.

General uses:

- **In the retail food industry:** Sauces, canned vegetables, fizzy drinks, etc.
- **In restaurants:** Found in products made from in-house fermentation.

Lactitol (E-966)

additives—sweeteners
additives—humectants

What is it? An artificial product from the polyol group, used as a sweetening and moisturizing additive.

Where does it come from? How is it obtained? From lactose.

Form: Powder.

Additional information:

- Sweetness: 0.3 times that of sugar (sucrose).
- Like all polyols, it may have a laxative effect in concentrations higher than 60 g/kg.

General uses:

- **In the retail food industry:** As a bulking agent for confectionery, caramels, chewing gums, jams; as a sweetener in low-calorie products and in chocolate; used for texture preservation and to reduce the freezing point of water solutions (food antifreeze).
- **In restaurants:** Used in frozen and semi-frozen desserts.

Lactose

food composition—carbohydrates

What is it? A sugar present in milk and dairy products.

Where does it come from? How is it obtained? It is extracted from milk.

Form: Powder.

Additional information:

- It is a natural disaccharide that is obtained basically from whey in the manufacture of cheeses.
- Its sweetening power is 0.2–0.6 times that of sucrose (sugar).
- Its molecule is formed by a glycosidic bond between a glucose and galactose molecule.

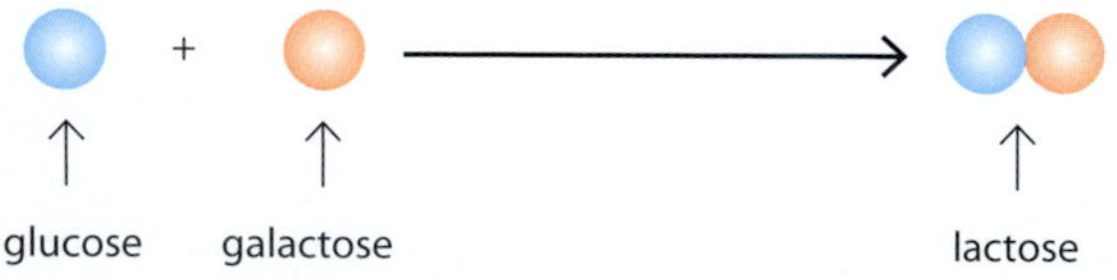
+
glucose
galactose
lactose

- Colorless and elastic toffees are obtained by substituting milk solids with lactose.
- People with lactose intolerance cannot tolerate large quantities of milk. Although some people are intolerant of even small quantities of milk, they can tolerate cheese and yogurt.

General uses:

- **In the retail food industry:** Caramels and sweet dairy products. It triggers the maturation of meat, is a bulking agent, etc.
- **In restaurants:** In experimentation.

Lambda (E-407)

additives—thickening agents

What is it? A fibrous carbohydrate from carrageenan that is used as a thickening additive. It has the properties of a hydrocolloid.

Where does it come from? How is it obtained? It is extracted from red algae. *See* Carrageenans.

Form: Powder.

Additional information:

- Unlike other carrageenans (kappa and iota) it does not have gelling effects; rather, it is a thickening product.
- Its thickening power is lower than that of xanthan, carobin, and guar, but it can be handled in hot or cold preparations, as well as allowing vigorous mechanical treatment (shaking, etc.).

General uses:

- **In the retail food industry:** Dairy desserts, creams and ice creams, fruit juices, processed meat, etc.
- **In restaurants:** In experimentation.

Lecithin (E-322)

additives—emulsifiers
additives—antioxidants

What is it? A natural additive from the phospholipids group that is used as an emulsifier and an antioxidant.

Where does it come from? How is it obtained? It is extracted from egg yolk or obtained by refining soy or sunflower oil.

Form: Powder, granulated, or liquid (in different concentrations).

Additional information:

- It is the natural emulsifying agent in egg yolk (between 30 and 35%) and in soy, and is present in other food products.
- It was discovered in eggs at the end of the 19th century.
- It is found in all living organisms, forming part of systems as important as the nervous system, including the brain.
- Vitamin E, with its antioxidant effects, is also obtained when lecithin is extracted from soy and sunflower oils.
- Lecithin is useful in the prevention of arteriosclerosis; it may also alleviate the effects of menopause.
- Antioxidant characteristics are attributed to it because of its sequestrant power (especially of iron).

Air bubbles obtained thanks to lecithin.

General uses:

- **In the retail food industry:** In animal or vegetable oils and fats. In the preparation of chocolates, salad products, milk and other dairy product derivatives, breads and pastries, etc.
    - Its main function is as an emulsifier, but it has other important uses: it reduces the high viscosity of (or makes fluid) chocolate paste during preparation, or else it is used as a humectant, so that very fine powder (for example, cocoa powder) can be rapidly dispersed in a liquid without forming lumps.
    - It also prevents fat bloom (the whitening of the surface of chocolate due to the migration of fat toward the exterior).
- **In restaurants:** It produces air bubbles (when added to liquids and mechanically shaken, it produces bubbles similar to soap bubbles) as well as in dressings and sauces to create a smooth and creamy mouthful. In experimentation for other uses.

Quantity and instructions for use:

- **Maximum/minimum quantity:** In animal and vegetable oils and fats without emulsifying (except virgin and other olive oils): 30 g/L. QS (minimum quantity required to obtain desired effect) in other applications.
- **Basic quantity for cooking:** From 0.3%; in other words, 0.3 g per 100 g of the liquid to be emulsified (3 g per kg).
- **Instructions for use:** It is mixed by shaking; it has an effect in both cold and hot preparations.

Levulose

See **Fructose**

Lime

mineral products

What is it? A mineral salt with the scientific name calcium oxide.

Where does it come from? How is it obtained? From calcareous rocks heated in special ovens.

Additional information:

- It is a white powder with a very high melting point (2,580°C).
- It is very hygroscopic and so transforms into a strong alkali, calcium hydroxide.
- It is used to obtain products such as calcium chloride.

- It is used to neutralize acidic soils in agriculture.
- Limewater may be used to help digestion and in fact is used in children's medicines to combat diarrhea and help milk digestion.

Lipids

food composition—lipids

What are they? They are biochemical products that are insoluble in water (hydrophobes or lipophiles). Any of a group of organic compounds including the fats, oils, waxes, sterols, nucleic acids, and triglycerides. Lipids are characterized by being insoluble in water(hydrophobes or lipophiles), and account for most of the fat present in the human body. They are, however, soluble in nonpolar organic solvents.

Additional information:

- This is the scientific name given to fats, oils, and other products with similar properties.
- They mainly act as energy reserves in the organism.

Lipophile

See **Hydrophobe**

Lipophobe

See **Hydrophile**

Liquid nitrogen (E-941)

additives—freezing agents

What is it? An element that is a gas above –196°C and is brought down to temperatures of between –196°C and –210°C in order to maintain a liquid state.

Additional information:

- If its temperature goes below –210°C, it becomes a solid.
- It has helped to develop one of the most important techniques in the world of frozen foods, due to its quick-freeze properties.
- It should be maintained in isolated containers and has a limited duration because it converts progressively into gas.

General uses:

- **In the retail food industry:** Quick-freeze or as a liquid cooling agent for special freezers.

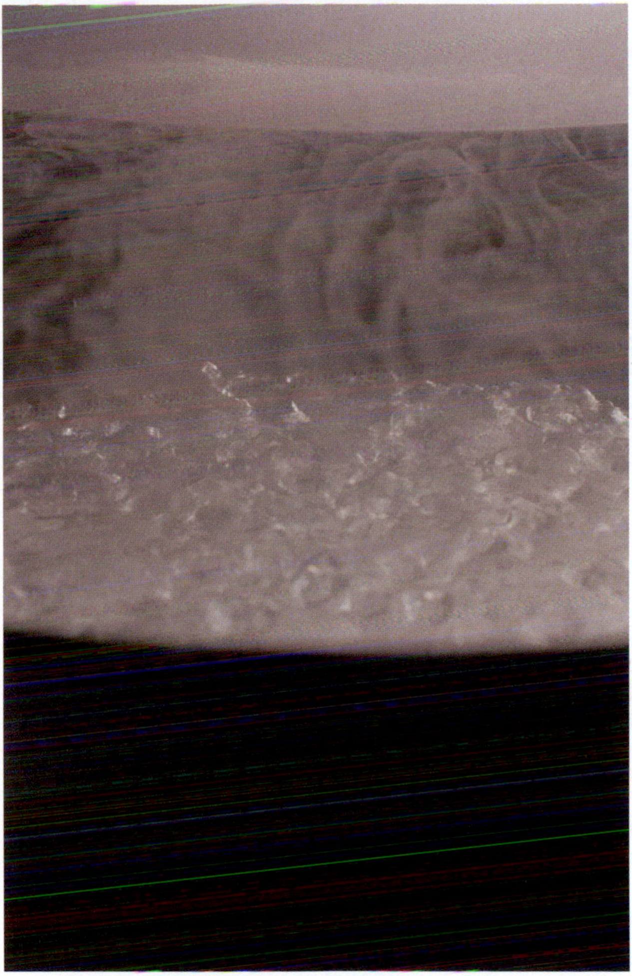

In restaurants: It enables sorbets and ice creams to be prepared and it is a cold "cooking" element for preparations such as mousses, foams, purées, jellies, etc.

Litmus paper

technology—devices

What is it? A strip of paper used to analyze the composition or characteristics of a foodstuff.

How does it work? By the action of substances impregnated on paper that normally adopts different colors, depending on whether there is a proportion of a determined component.

Additional information:

- There are many different types of litmus paper. It is normally used to detect:
  - Acidity (pH litmus paper).
  - Hardness (it indicates the presence of calcium and magnesium).

General uses:

- **In the retail food industry:** Analysis of products and quality control.
- **In restaurants:** In house-made sanitizing solutions.

LM Pectin (E-440)

food composition—carbohydrates
additives—gelling agents

What is it? A soluble, fibrous-type complex carbohydrate (polysaccharide) that is used as a gelling additive. It has the properties of a hydrocolloid.

Where does it come from? How is it obtained? It is present in the cell wall of vegetables (protopectin).

Form: Powder.

Additional information:

- Classification:
  - Amidated LM pectins (LA).
  - Conventional LM pectins (LM).
- It gives a thermoreversible gel, i.e., it is a gel or not, depending on the temperature (unlike HM pectin).
- It gels in the presence of calcium and other similar salts and does not need sugar or acid in order to act (unlike HM pectin). In some cases (dairy products and even certain fruits), the calcium that is already present in the foodstuffs is sufficient.

General uses:

- **In the retail food industry:** Jams and marmalades, canned vegetables, baking, juice derivatives, and fruit juice.
- **In restaurants:** In gels, jellies, and experimentation.

See HM pectin.

Lyophilization (freeze drying)

physical or chemical processes

What is it? It is a technique that uses dehydration by sublimation (direct conversion from solid to gas), which is carried out in a container where there is a vacuum and low temperature.

Additional information:

- Low temperatures prevent product alterations and losses of volatile components; this is useful in nutrition to preserve tastes and aromas.
- Lyophilizers are expensive and the technique therefore does not tend to be used in cooking, but it is used in the retail food industry when aromas need to be preserved and when other methods are unadvisable. Widely used in the pharmaceutical industry.

General uses:

- **In the retail food industry:** In fruit, aromatic herbs, spices, coffee, vegetables, milk, and in preparations for expeditions where weight is an important consideration. This is the procedure most commonly used for human space travel.
- **In restaurants:** In experimentation. However, more frequently previously lyophilized ingredients will be used.

Lyophilizer

technology—devices

What is it? A device that enables products and preparations to be dehydrated by sublimation (direct conversion from solid to gas).

How does it work? The product is placed in the device and the water is sublimated at low temperatures (–50°C to –80°C) in a nearly total vacuum, without resulting in product flavor loss.

Additional information:

- A lyophilizer is composed of
  - A refrigerating compressor
  - A vacuum pump
  - A condenser
- A lyophilizer is a totally automated control system.

Besides its basic function (lyophilization), it can also reduce volume (by partially extracting water from a foodstuff or a mixture of foodstuffs).

See Lyophilization.

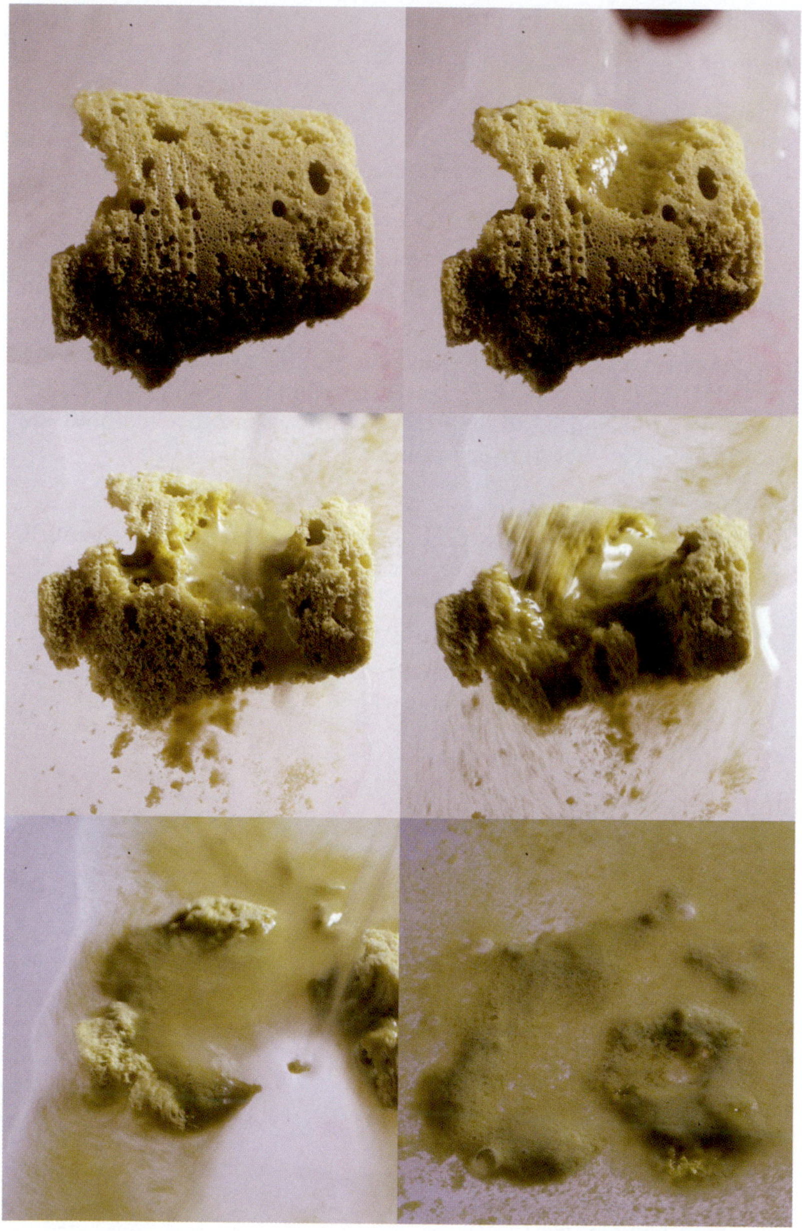

Water is added to lyophilized pistachio foam in order to rehydrate it.

M

Magnesium
food composition—minerals

What is it? A chemical element that is present in mineral salts and used as a food complement.

Where does it come from? Mineral salts.

Additional information:

- It is indispensable for living organisms. There is approximately 0.2 g per kg in the human organism.
- It acts when enzymes generate energetic products for the organism. Bones are made of magnesium and calcium (mostly calcium).
- It is consumed only through foodstuffs or supplements, basically through pulses and nuts: chickpeas (108 mg per 100 g), almonds (170 mg per 100 g), etc.

Magnetic mixer
technology—devices

What is it? A device that allows a liquid to be blended regularly and constantly. It provides total control over the mixing speed of the liquid as well as the duration.

How does it work? A small magnet is placed in a bowl with the liquid that needs blending. A revolving magnet situated inside the device generates a magnetic force that causes the magnet in the liquid to revolve and the liquid in the bowl to be blended.

Additional information:

There is a thermomagnetic mixer version where the liquid product may also be heated.

General uses:

- **In the retail food industry:** In research and quality control processes.
- **In restaurants**: In experimentation.

Maillard reaction

physical or chemical processes

What is it? A highly complex set of chemical reactions between amino acids and carbohydrates that occurs when certain foodstuffs are submitted to high temperatures (during grilling, roasting, barbecuing, stewing, etc.), lending them a brown color and a characteristic taste.

How is it produced? When a foodstuff is heated, an amino acid (protein component) and a carbohydrate (glucose, fructose, etc.) react.

Additional information:

- Despite the reaction beginning at 30–40°C, it is apparent only at 130°C.
- These reactions can cause dark (earthy) colorations and flavors related to the cooking process. Likewise, small quantities of carcinogenic substances, some of which have an unpleasant flavor, are formed, especially at high temperatures.
- Coloration is sometimes desirable (bread crust, biscuits, beer) and sometimes undesirable (milk and concentrated fruit juices). The flavors can be well defined (roast beef, barbecued sardines, cocoa, etc.) and this is why they are frequently used by the retail food industry.
- The intensity of the reaction depends on the temperature, the quantity of reducing carbohydrates (basically monosaccharides and disaccharides) and the quantity of proteins, but also on other factors such as pressure, humidity, other components, etc.
- The reaction was named after Louis-Camille Maillard, who discovered and defined these reactions in 1912.

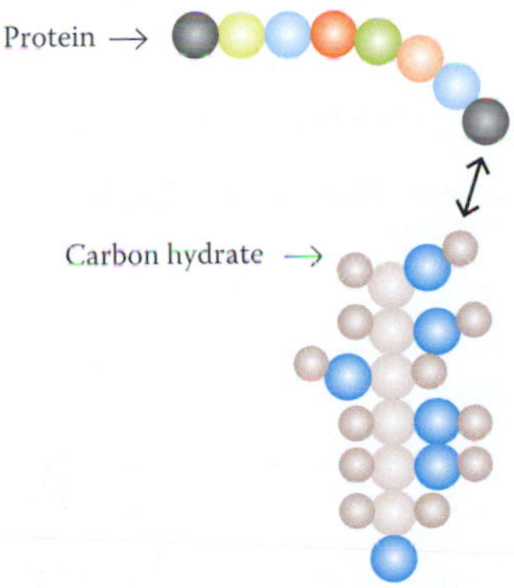
Protein
Carbon hydrate

At high temperatures, carbohydrates react with proteins and form characteristic compounds. In meat, they are responsible for the brownish crust that maintains the juices if properly cooked.

Malic acid (E-296)

additives—acidity regulators
food composition—acids

What is it? An acid that is used as an acidity regulator.

Where does it come from? How is it obtained? From fruit, especially apples.

Form: Powder or crystallized.

Additional information:

- When fruit ripens, malic acid is normally converted into tartaric acid.
- It is used as an acidifier, for example, in the preparation of fruit pastes.
- Its acidity can be used to stimulate the taste buds, making them more sensitive to other tastes.

General uses:

- **In the retail food industry:** Jams, jellies, soft drinks, canned fruits, and vegetables, etc.
- **In restaurants:** In experimentation.

Maltitol (E-965)

additives—sweeteners
additives—humectants

What is it? An artificial additive from the polyol group that is used as a sweetening and moisturizing product.

Where does it come from? How is it obtained? It is obtained in the chemical industry by synthesis of maltose, which is taken from starch.

Form: Powder and in solution (maltitol syrup).

Additional information:

Sweetening power: 0.8 times that of sugar (sucrose).

General uses:

- **In the retail food industry:** Chocolate, caramels, chewing gum, biscuits (in dough as a sucrose substitute), etc.
- **In restaurants:** Used in pastry and confectionary development.

Maltodextrin

food composition—carbohydrates

What is it? A carbohydrate that is formed when starch chains break down.

Where does it come from? How is it obtained? By physicochemical treatment of starches (wheat, corn, tapioca, etc.).

Form: Powder.

Additional information:

- According to regulations, carbohydrates are called maltodextrines when they have a dextrose equivalence below 20. There are, therefore, maltodextrins with varying DEs; for example, 18 DE, 16 DE, etc.
- It has a slightly sweet taste because of the minimal presence of sugars (monosaccharides and disaccharides) in its composition (0.1–0.2 times the sweetening power of sugar). For example: maltodextrine DE 15 means that 15% is composed of reducing sugars, basically of monosaccharides and disaccharides.

General uses:

- **In the retail food industry:** As a bulking agent (it increases the volume of the foodstuff without noticeably varying the organoleptic elements).
- **In restaurants:** In experimentation.

Maltose

food composition—carbohydrates

What is it? A disaccharide carbohydrate that is used as a sweetener.

Where does it come from? How is it obtained? By physicochemical treatment of starches (wheat, corn, tapioca, etc.).

Form: Powder.

Additional information:

- Sweetening power: 0.3–0.6 times that of sugar (sucrose).
- It is formed by the bonding of two glucose molecules.

+
glucose
glucose
maltose

- It is not manufactured for sale on its own, rather as a part of glucose syrups.

General uses:

- **In the retail food industry:** Normally in combination with other carbohydrates in glucose syrups that are used for desserts.
- **In restaurants:** No known use other than as a component of glucose syrup.

Maltotriose

food composition—carbohydrates

What is it? A carbohydrate formed by three glucose molecules (trisaccharide).

Where does it come from? How is it obtained? Through treatment of starch.

Additional information:

It is a component of glucose syrups. *See* Glucose syrup.

Mannitol (E-421)

additives—sweeteners

What is it? An artificial product from the polyol group that is used as a sweetening additive.

Where does it come from? How is it obtained? It is obtained in the chemical industry by synthesis from fructose or invert sugar (a mixture of glucose and fructose).

Form: Powder.

Additional information:

- It is similar to sorbitol. In fact, they are obtained together.
- Sweetening power: 0.6 times that of sugar (sucrose).

General uses:

- **In the retail food industry:** In hard caramels and candies, chewing gum, it reduces "stickiness"; also used as an anti-caking agent due to its reduced tendency to retain water.
- **In restaurants:** No known use.

Mannose

food composition—carbohydrates

What is it? A simple carbohydrate (monosaccharide) that forms part of some hydrocolloids, for example, alginates, carob gum, xanthan gum, etc.

Matter

scientific concepts

What is it? Everything that physically constitutes the universe as we know it.

Additional information:

Matter is classified into **heterogeneous** matter (the different components can be distinguished visibly) and **homogeneous** matter (the components are not visibly apparent).

Classification of Matter		
Heterogeneous		
Homogeneous	Homogeneous mixtures	
	Pure substances	Chemical compounds
		Chemical elements

Measuring cylinder (graduated cylinder)

technology—devices

What is it? A utensil that is used to measure liquid volumes (wine, juice, etc.).

Additional information:

The most widely used measuring cylinders have capacities of 25, 50, 100, 250, 500 and 1,000 mL.

Melanin

food composition—pigments and other compounds

What is it? A dark pigment that is present in fruits and vegetables when they turn brown, and in some animals.

Additional information:

- It determines color and taste in squid and cuttlefish inks.
- The skin of mammals also produces this pigment as protection against solar radiation.

Melting salt

food concepts

What is it? The name given to different types of salt capable of reorganizing the structures of proteins contained in cheese.

Additional information:

- They allow cheeses to be melted and extended easily.
- Calcium acts as a "cement" in cheeses to unite the protein particles—very hard textures may be produced over time depending on the type of cheese. If this hard cheese is melted with salts (melting salts) that are capable of "sequestrating" the calcium, the protein relaxes and the cheese can be spread.
- The salts that are most commonly used to achieve this effect are phosphates and citrates. The name "melting salts" stems from the period when edible but very hard cheeses were used by melting them with salts. Nowadays, spreading cheese or fondue cheese is prepared by adding the salts at the beginning of the preparation.

Methylcellulose (MC) (E-461)

additives—thickening agents
additives—gelling agents

What is it? A fibrous carbohydrate that is used as a gelling and thickening additive. It has the properties of a hydrocolloid.

Where does it come from? How is it obtained? From the reaction that takes place when methyl groups are added to plant cellulose.

Form: Powder.

Additional information:

It is thermoreversible (gel–non-gel depending on the temperature) but it acts differently from other gels: jelled when hot (more than 50°C) and liquid when cold.

General uses:

- **In the retail food industry:** Crème caramel, puddings, stuffings, béchamel, pizza toppings, croquettes, etc.
- **In restaurants:** Used in sauces, dressings, and in making hot ice cream.

Cloud of passion fruit juice, made possible by methlycellulose.

Microcrystalline cellulose (MCC) (E-460)

additives—thickening agents
additives—stabilizers

What is it? A fibrous carbohydrate that is used as a thickening additive. It has the properties of a hydrocolloid.

Where does it come from? How is it obtained? From plant cellulose. It is a breaking up of wood or cotton pulp when their long-chain structures are partially broken down by physicochemical processes.

Form: Powder.

General uses:

- **In the retail food industry:** Used as a dietetic fiber, to prevent grated cheese from sticking, an aroma encapsulator due to its capacity to absorb essential oils, a suspending agent in milkshakes, a stabilizer in emulsions and foams.
- **In restaurants:** As an acti-caking agent and as a dry flavor carrier.

Microorganism (or microbe)

scientific concepts

What is it? A living being that is visible only under a microscope.

Additional information:

- Bacteria and fungi are the microorganisms that are most related to food.
- There are many species in each family; some are dangerous for food preservation and health, such as salmonella, and others are beneficial, such as the bacteria that are used to make yogurt, yeasts used to make breads and other fungi that cause special flavors in cheese (Roquefort) and sausages (mold that forms on the skin).

Microorganisms		
Bacteria	**Fermentation**	**Lactic bacteria:** Lactobacilli bacteria; they produce lactic acid.
		Acetic bacteria: Bacteria of the *Acetobacter* genre, which produce acetic acid.
	Obtaining products	*Xanthomonas campestris*, which produces the thickener xanthan gum.
		Sphingomonas elodea, which produces the gelling agent gellan gum.
Yeasts (fermentations)	*Saccharomyces cerevisiae*: some 20 different species. Depending on the species, used for fermentation of bread, beer, wine, etc. For example: *S. oviformis* for wine varieties.	
	Saccharomyces bayanus For example: *S. uvarum* for wine varieties.	
	Saccharomyces rosei For example: *Torulaspora delbrueckii* for wine varieties.	

Microwave
scientific concepts

What is it? Electromagnetic radiation, with a wavelength between infrared and radio waves, which can be used to heat polar molecules (for example, water).

Additional information:

- Microwaves produce a change of direction in an electrical field approximately 5 million times per second, causing the polar molecules to change direction. When these molecules move, generally water, internal frictions are generated that heat the substance, without exceeding 100°C (boiling point of water). However, when fats, sugars, and alcohols are heated, higher temperatures can be reached.

- The microwave principle is used in the kitchen in microwave ovens, where products can be heated and cooked as if they were being boiled; the water that the product or preparation contains is heated and cooked, but unless there are exceptional circumstances (high concentration of fats and sugars), the roast flavor normally achieved by cooking over a barbecue, under a grill, or in a conventional oven, etc., will not be obtained (Maillard reaction).
- Frozen products or products with little water are heated irregularly and sometimes part of a dish boils while the other part remains cold. The turntable rotates to reduce this problem.
- In the case of foodstuffs with low water content, microwaves can act on certain fat, sugar, or alcohol molecules, which can lead to high heat that can causes burns.
- The behavior of metals inside the microwave oven is unpredictable due to the properties of metals (conductivity, wave reflection). It is not recommended, therefore, to place metallic objects inside microwave ovens.

Mineral salts

food composition—minerals

What are they? Chemical products that are not formed by carbon structures and that form foodstuffs together with biochemical compounds and water. They are derived from the reaction of an acid with an alkali.

Additional information:

The most common in the retail food industry are calcium salts and potassium salts.

Minerals

scientific concepts
mineral products

What are they? Products that are not derived from living beings (as opposed to animal and vegetable products), despite the fact that they may be contained in them.

Additional information:

If the mineral products contained in foodstuffs are in the form of salts they are called mineral salts.

Modified products
scientific concepts

What are they? Natural products that have been modified structurally in order to acquire different properties.

Additional information:

For example: Modified starches. Native (natural) starches pose some problems, such as their difficult hydration and the retrogradation phenomenon. *See* Retrogradation. With gentle physical or chemical treatment a series of modified starches can be produced, each one with a particular property (instant hydration, no retrogradation, more or less viscous, resistant to enzymes, etc.).

Moisturizing agent
food concepts

What is it? A product that prevents the drying of foodstuffs by retaining water and counteracting the effect of a dry environment.

Additional information:

- Allows powder products to be quickly dispersed in liquid without forming lumps.
- In industrial food products, some phosphates such as E-343 or else malates E-350 to 352 (derived from malic acid) are used as humectants.
- Some polyol-type sweeteners are also humectants: lactitol, maltitol, and sorbitol. These products can be used in kitchens to avoid the drying out of products such as croquettes, breads, etc.
- To avoid the drying out of sliced or other breads, monoglycerides and diglycerides (E-471) are used. They surround the starch amylose chains and prevent the release of water (they prevent starch from retrograding).

Molecular gastronomy
scientific concepts

What is it? A scientific term, coined in 1988 by Nicolas Kurti and Hervé This to denote the study of scientific phenomena in cooking. However, this debate between science and cooking did not become a reality until the 21st century.

Molecule
scientific concepts

What is it? A specific grouping of atoms that are linked by chemical bonds of varying stability.

Additional information:

Molecules can also be formed by the bonding of simpler molecules. They have specific properties that are different from those of the atoms or simple molecules that they are composed of.

For example:

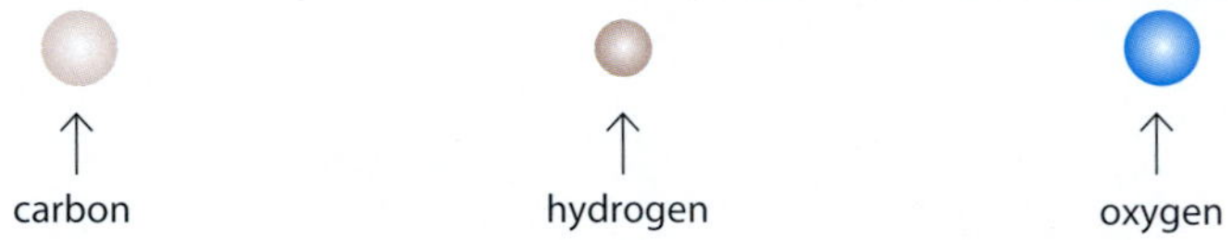
carbon
hydrogen
oxygen

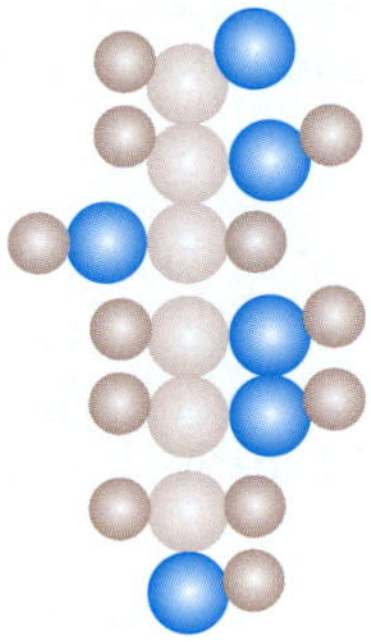

Glucose
(6 carbon atoms, 6 oxygen, and 12 hydrogen)

Monoglycerides and diglycerides (E-471)

additives—emulsifiers

What are they? Additives derived from the reaction between glycerine and fatty acids, used as emulsifiers.

Where do they come from? How are they obtained? From a chemical reaction called esterification, from glycerine and the fatty acids of fats.

Form: Powder, flakes, or liquid.

Additional information:

- The most commonly used monoglycerides and diglycerides have a low HLB (between 3 and 4), which means that they are used to prepare water–oil emulsions (W/O); that is, the emulsifier has an affinity for oil. This means that they must first be broken down by the fatty element and have the watery element added at the end.

General uses:

- **In the retail food industry:** Ice creams, oils (animal or vegetable), quick-cook rice, cocoa and chocolate, milk and dairy products (pasteurized whole cream), bread, fresh pasta, preserves, spreadable jams and jellies that are prepared from fruit, including low-energy products, juices, etc.
- **In restaurants:** Used in ice cream production.

Monosaccharides

food composition—carbohydrates

What are they? They are simple molecule carbohydrates; they have a sweetening power and form part of sugars.

Additional information:

- From the gastronomic point of view, the most common monosaccharides are glucose and fructose, which can be found in nature in honey and fruit. Their most important properties are
  - They are highly soluble in water.
  - They have a sweet taste. This property means they are frequently used in gastronomy, above all in the world of desserts and sweets, from jelly candies to cakes.
  - They act as preservatives when sugar proportions are high.

See Carbohydrates.

To prepare, olive oil, caramel, isomaltitol, glucose, and sucrose ester are mixed together and cooked at 160°C. While the caramel is being cooked, monoglyceride should be dissolved with virgin olive oil at 50°C. When the caramel reaches 160°C the olive oil is drizzled in and blended in with a spatula. Once the caramel has absorbed all the oil, it is stretched out on a piece of sulphurized paper. Finally, a knife is used to mark 5 by 5 cm squares.

Mouthfeels
organoleptic perceptions

What are they? They are defined and recognizable taste notes that do not include the basic tastes. Some types are sour, balsamic, astringent, rancid, iodized, smoked, spicy, and metallic. A product's physical and chemical interaction in the mouth, an aspect of *food rheology*. It is a concept used in many areas related to the testing and evaluating of foodstuffs, such as *wine tasting* and *rheology*. It is evaluated from initial perception on the *palate*, to first bite, through *mastication* to *swallowing* and *aftertaste*. In wine tasting, for example, mouthfeel is usually used with a modifier (big, sweet, tannic, chewy, etc.) to the general sensation of the wine in the mouth. Some people, however, still use the traditional term, "texture." Mouthfeel is often related to a product's *water activity*, hard or crisp products have lower water activities and soft products have intermediate to high water activities. Some types are hard or soft, rough, silky, sour, balsamic, astringent, rancid, iodized, smoked, spicy, metallic, etc.

Mucilage
food concepts

What is it? Its name is synonymous with hydrocolloid and gum. In some foodstuffs (for example, the cocoa bean) the part of the bean capable of absorbing water is referred to as the mucilaginous layer.

Myoglobin
food composition—proteins

What is it? A red protein (myoglobin with oxygen) similar to a subunit of hemoglobin. It is found in the muscles of vertebrate animals and is responsible for storing and carrying oxygen.

Additional information:

- The prolonged oxidation of myoglobin produces brown-colored metamyoglobin and indicates that a certain length of time has passed; the smell should indicate whether decomposition has begun.
- Some cooking methods for meat (charcoal, wood, or gas) can give rise to a pinkish color.

On creating a vacuum, for example for a piece of veal, the myoglobin loses oxygen and acquires a brown, almost violet hue. When the product comes into contact with oxygen again, it becomes red.

N

Naringin
food composition—pigments and other compounds

What is it? A natural flavonoid-type substance found in grapefruit and other citrus fruit (in the pith of oranges and lemons).

Where does it come from? How is it obtained? It is extracted from citrus fruit.

Form: Powder.

Additional information:

- It has a bitter taste and is used in preserves (bitter orange), drinks (bitters), etc.
- It is considered nutraceutical because of its quality as a flavonoid. *See* Nutraceutical products.

General uses:

- **In the retail food industry:** Used as bitters in drinks.
- **In restaurants:** Used as bitters in drinks.

Natural products
scientific concepts

What are they? Products that are obtained from nature and consumed directly or submitted to a slight process in order to be adapted for consumption. The USDA's current definition says that a product labeled as natural should not contain any artificial flavor, coloring or chemical preservative.

Additional information:

- They can be classified into groups:
    - Of animal origin: Veal collagen, cow's milk, prawns, etc.

- Of vegetable origin: Apples, agar agar (a product extracted from algae).
- Minerals: Salt (sodium chloride), potassic salt (potassium chloride), water, etc.

- "Slightly processed natural products" are products obtained from nature, but require a slight physical or chemical treatment before consumption, normally related to extraction techniques. One example is skimmed or semi-skimmed milk, which has had its fat totally or partially removed by energy centrifuge.
- Another example is refined white sugar used for normal consumption. Physical processes such as extractions and filtrations are used to obtain it from sugar cane or sugar beets, but complementary chemical treatments are also required.

Neohesperidin-DC (E-959)

additives—sweeteners

What is it? An artificial additive derived from sugar-type carbohydrates and used as a sweetener.

Where does it come from? How is it obtained? Industrial chemical synthesis of bitter orange rinds in the initial stages of formation.

Form: Powder.

Additional information:

- Its abbreviation is NHDC.
- Some types of bitter orange have a natural substance, hesperidin, with a bitter flavor; it is modified to obtain the neohesperidin that is 600 times sweeter than sugar.
- Its high price means that it is used in combination with other sweeteners, or in tiny doses as a flavor enhancer.

General uses:

- **In the retail food industry:** Confectionery, drinks, chewing gum, jams, and as a sweetener for diabetics etc.
- **In restaurants:** No known use.

Neotame
additives—sweeteners

What is it? A very powerful sweetener 8,000 times sweeter than sugar (sucrose). It was developed by the chemical-pharmaceutical company Monsanto and is in the regulation and application phase.

Neutralization
physical or chemical processes

What is it? A process or chemical reaction during which an acid and an alkali react with each other until the properties of both disappear.

Additional information:

- In these reactions the acid and the alkali disappear and a type of mineral salt and water becomes available:

  acid + alkali = mineral salt + water

- An example is the neutralization of stomach acids with bicarbonate.

Nitrate salt (E-252)
additives—preservatives

What is it? A preservative with the formula of potassium nitrate.

General uses:

- **In the retail food industry:** Processed meats such as sausages, hot dogs, bacon, etc.
- **In restaurants:** In-house processed meats.

See Nitrates and Nitrites.

Nitrates and nitrites (E-249 to E-252)
additives—preservatives

What are they? Salts that contain nitrogen and oxygen, and are used as preservatives.

Where do they come from? How are they obtained? They are found in natural deposits as well as in some vegetables (spinach, broccoli, red pepper). They are known as Chile nitrates because of their origin.

Form: Powder.

Additional information:

- From ancient times brine has been prepared to pickle meat, fish, and cheese with "nitrate salt" (potassium nitrate), which contributes to the preservation of foodstuffs and, in the case of meat, improves its color.
- Nitrates alone are not effective. The preservative property comes mostly from nitrites on the bacteria found naturally in meats and cheese. As preservatives, the combination is used because nitrates will convert into nitrites in the food product. If only nitrites are used, then the needed levels might exceed regulations and if only nitrates are used the conversion reaction may take too long.
- Nitrites are currently the only substances capable of preventing botulism intoxication (they prevent the growth of the bacterium *Clostridium botulinum,* which is still fatal in badly sterilized preserves.
- In certain circumstances nitrites can produce carcinogenic products. To avoid this problem they are now always added together with vitamin C.

General uses:

- **In the retail food industry:** Processed meats such as sausages, hot dogs, bacon, etc.
- **In restaurants:** In-house processed meats.

Nitrogen (E-941)

additives—preservatives
additives—gases

What is it? An element. It is a gas component of air (78%) that is used as a preservative.

General uses:

- **In the retail food industry:** It is used in what are known as modified atmospheres that enable foodstuffs to be preserved with no contact with oxygen. It is also used to maintain pressure in tetra briks, thus preventing their deformation during vacuum packaging.

In restaurants: Found in cartridges for making aerated foods such as foams. *See* Nitrogen protoxide.

Nitrogen protoxide (E-942)

additives—gases

What is it? An inorganic product formed by nitrogen and oxygen (N_2O) and used as an additive for applications such as packaging gas.

Where does it come from? How is it obtained? By chemical reactions of nitrogen products with mineral origin.

Form: Compressed gas.

General uses:

- **In the retail food industry:** As a propellant inside airtight containers, for example, cream spray.
- **In restaurants:** In capsules for siphons.

Nutraceutical products

food concepts

What are they? Intermediate food products that are added to foodstuffs, supposedly or actually to contribute to maintaining the health of those who consume them.

For example: Milk with omega-3 fatty acids to help reduce cholesterol.

Main Nutraceutical Products	
Oligosaccharides	Choline
Unsaturated fatty acids	Phospholipids
Peptides	Lactic bacteria
Glucosides, isoprenoids	Vitamins
Polyphenols	Minerals

Nutrient
scientific concepts

What is it? Any substance found in foodstuffs that is useful for organic metabolism and belonging to the groups generically named proteins, carbohydrates, lipids, minerals, vitamins, and water. Indispensable for good health.

Additional information:

- Macronutrients are the nutrients that are found in large quantities in foodstuffs (proteins, lipids, and carbohydrates).
- Micronutrients are found in very small quantities (vitamins, minerals).

Nutrition
scientific concepts

What is it? A basic function for living organisms to ensure development and optimum growth.

O

Oils

food composition—lipids

What are they?

- **Scientifically:** Specific types of triglycerides that are liquid at room temperature.
- **Gastronomically:** Fatty substances with a fluid texture at room temperature. Usually of vegetable origin although animal oils also exist.

Additional information:

- They are related to unsaturated fatty acids because they have low fusion points. This means that they may remain liquid at room temperature.
- The unsaturated fatty acids in their composition mean that they are healthy, and it has even been said that they are nutraceutical.
- Natural oils are a blend of triglycerides.

Oligoelement

scientific concepts

What is it? A chemical element, required by the organism but in very small quantities.

Additional information:

Normally it is required in the form of the mineral salt of an element; for example, fluorine, nickel, copper, iodine, etc.

Oligosaccharides

food composition—carbohydrates

What are they? Carbohydrates formed by two or more different monosaccharides.

Additional information:

- The most important are sucrose, raffinose and stachyose. *See* Sucrose, Raffinose.
- They are used in the retail food industry as nutraceutical products.

Omega-3 fatty acid

food composition—lipids

What are they? Lipids used as nutraceutical products.

Additional information:

- They are found in foods such as oily fish, pulses, nuts, etc. Nowadays, due to their preventive effects in coronary diseases (heart attacks, angina, etc.), they are isolated and added to margarine, milk, etc., as products that reduce cholesterol.
- The name is derived from the last letter of the Greek alphabet "omega," indicating with the number (3) the unsaturated position in the fat molecule, which in this case is three carbons from the end.

Organic

See **Chemical compound**

Organoleptic

scientific concepts

What is it? It is the capacity of a foodstuff to produce an effect on the senses (sight, smell, touch, taste, and hearing), so that it can be perceived, distinguished, and appreciated.

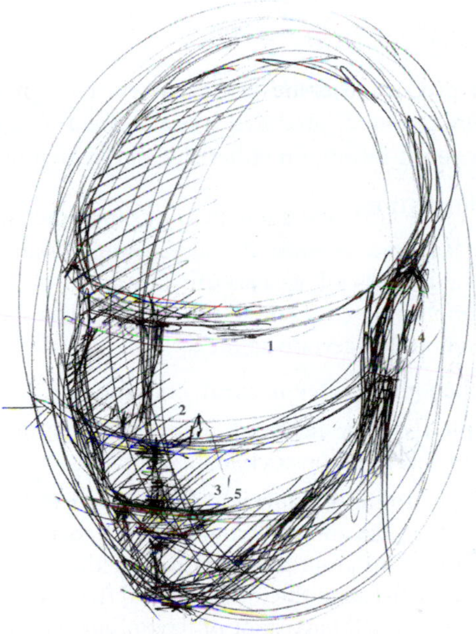
1
2
3
4
5

Role of the Senses When Eating	
Sight	**Taste**
shapes and proportions	**Perception of primary tastes**
colors	sweet
layout on the plate	savory
identification of the product	acidic
identification of the style	bitter
"reading" a dish, etc.	umami (*See* Umami)
Smell	**Perception of mouthfeels**
smell (products, preparations, condiments)	sour
Touch	balsamic
temperatures	rancid
textures	iodized
Hearing	spicy, etc.
perception of sounds (crunch, etc.)	**Flavor of foodstuffs ("gene")**

Osmosis

physical or chemical processes

What is it? A process whereby water passes through a semipermeable membrane from a more diluted to a more concentrated solution, thereby balancing the concentrations on either side of the membrane.

Additional information:

- The cell membranes of animal and vegetable tissues are naturally semipermeable. This allows certain substances to be continuously absorbed or eliminated through the membrane, especially the solvent, but not others (certain solutes).
- Osmotic phenomena often occur in cooking:
  - **Example 1:** When meat, fish, or other foodstuffs are cooked in a watery substance (a stock, for example):
    - If salt (sodium chloride) is not added to the cooking liquid, the salts and the aromatic substances contained in the foodstuff migrate to the liquid medium to achieve a salt balance between the foodstuff and the stock. Consequently, the stock will have a lot of flavor, but the foodstuff will be more insipid.
    - If salt is added before cooking, the mineral salts and the aromatic substances contained in the foodstuff do not migrate to the stock or sauce because the elements are already balanced. Consequently, the foodstuff will be flavorsome and the stock bland.
  - **Example 2:** The osmotic effect is a traditional method for preserving certain foodstuffs. If abundant quantities of salt or sugar are added to a foodstuff the water contained in the microbes will migrate through the delicate membrane, resulting in the collapse and eventual destruction of the microbes. Salted cod and anchovies and cured ham are examples of the salt effect. A jam with more than 50% sugar needs no more protection against microbes.
  - **Example 3:** When fruit is placed in water, osmosis will cause the water to enter the fruit cells in order to balance sugar concentrations; the cells will accumulate water and burst. On the other hand, when too much sugar is placed in the water, the fruit will wrinkle. It is therefore important to find the optimum concentration for the preparation of fruit in syrup.
  - **Example 4:** Canarian wrinkly potatoes—*Papas arrugadas* in Spanish (salty potatoes boiled in their skins). These famous potatoes, cultivated in the volcanic soil of the Canary Islands, are cooked to optimize the osmotic effect. The potatoes are

barely covered by very salty water and boiled. To balance the salt concentration, the potato juices migrate. Consequently, the potatoes wrinkle and as the surface water evaporates a layer of crystalline salt covers the skins.

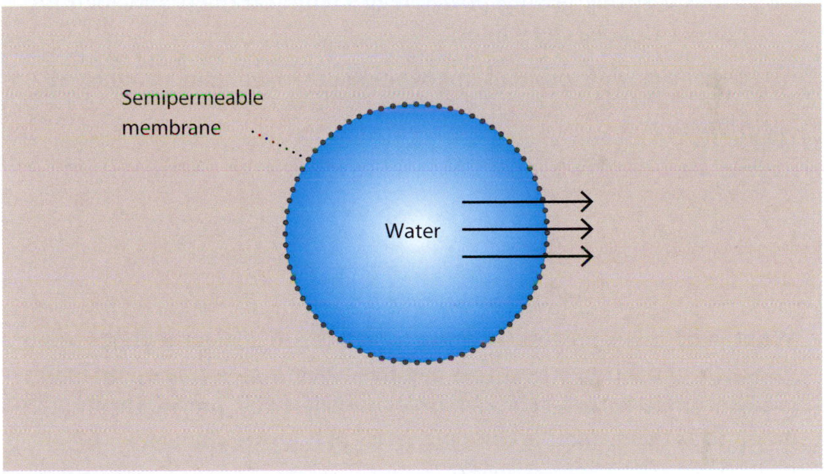
Semipermeable membrane
Water

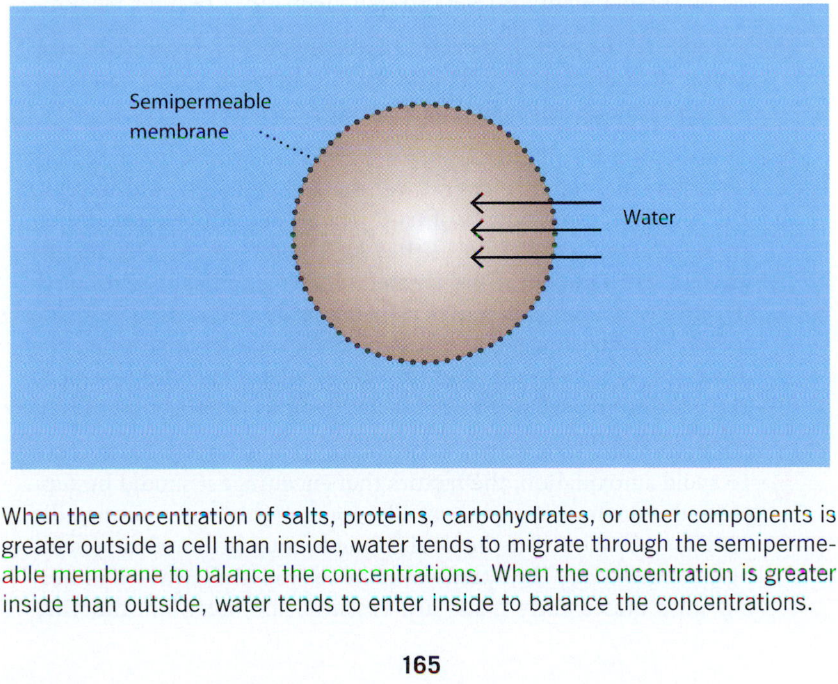
Semipermeable membrane
Water

When the concentration of salts, proteins, carbohydrates, or other components is greater outside a cell than inside, water tends to migrate through the semipermeable membrane to balance the concentrations. When the concentration is greater inside than outside, water tends to enter inside to balance the concentrations.

Ovalbumin
food composition—proteins

What is it? The main protein in egg whites; it contains phosphorous and mannose-type carbohydrates.

Additional information:

- It has a strong gelling power (egg crème caramel) and foaming power (whipped egg whites).
- Other egg white proteins are ovoglobulin, ovomucin, ovomucoid.

See Albumins.

Oxidation of foodstuffs
physical or chemical processes

What is it? It is a process whereby foodstuffs in contact with air age and lose their initial properties. This happens when a molecule or ion (atom or group of atoms with electrical charge) loses electrons and another molecule or ion wins them, and the characteristics of both are changed.

Additional information:

- Oxygen does not always cause the oxidation phenomena (for example, the oxidation of iron salts so that green olives become black).
- In cooking, the most important oxidizing phenomenon is the oxidation of triglycerides (oils and fats), otherwise called rancidity development, as well as that of fruit and vegetables.
- Autoxidation is a highly complex process in which many factors can intervene: light, temperature, metals, pigments and oxygen. The "weakest" components of triglycerides regarding autoxidation are unsaturated fatty acids, as they have points (the double bonds) where a free radical can be formed. The free radical is extremely sensitive to oxygen; it captures it from the air and forms a peroxide radical that attacks a neighboring molecule in order to reinitiate the process. It is a chain reaction that accumulates harmful products. There are no strange smells and tastes to indicate danger (as in rancidity development).
- To avoid autoxidation, the factors that encourage it should be suppressed: store in a dark place at a low temperature, use nonmetallic containers or eliminate the double bonds of unsaturated fatty acids by hydrogenation. The latter would be the most efficient technique, but unsaturated fatty acids are essential for nutrition and they

should be preserved. In extreme cases, the triglyceride should be protected with an antioxidant.

- Autoxidation causes the vitamins A, D, and E and the essential fatty acids to be lost and nutritional value to decrease.

Oxygen (E-948)

additives—gases

What is it? An element. It is a gas component of air (21%) and is responsible for food oxidations. As an additive (E-948), it is used to cause planned and controlled oxidations.

General uses:

- **In the retail food industry:** It is used for oxidations that cause fats to age, for example, in the meat industry. It is also used as a reagent to obtain iron oxides (E-172).
- **In restaurants:** Used in the dry aging of meats and as part of an aeration ingredient when oxidation will be an issue.

P

Papain

See **Enzymes**

Pasteurization

physical or chemical processes

What is it? It is a process whereby a liquid is submitted to the action of temperature in order to eliminate the pathogenic microorganisms or to considerably reduce their quantity.

Types of pasteurization:

- **Pasteurization LTLT:** At low temperatures and for long periods of time; for example, at 60–65°C for 20–40 minutes.
- **Pasteurization HTST:** At high temperatures and for short periods of time; for example, at 70–80°C for 15–30 seconds.
- **Pasteurization UHT:** At high temperatures and for very short periods of time; for example, at 135–140°C for 1–5 seconds.

Additional information:

- It is a widely used process in the juice and dairy industries.
- The core of the product must be reached if the pasteurization process is to be completed.
- A pasteurized product has a relatively short expiration date because only the pathogenic microorganisms are eliminated in order to preserve the organoleptic characteristics of the foodstuff.

Peptides
scientific concepts

What are they? The generic name for short-chain compounds formed by the bonding of two or more amino acids.

Additional information:

- They are called dipeptides (two amino acids), tripeptides (three) and polypeptides (many). They have the typical protein structure.
- They have little gastronomic value although they are used in the retail food industry as nutraceutical products.

For example:

- Aspartame is a bond between two amino acids: aspartic acid and phenylalanine.
- It is considered to be 180–200 times sweeter than the sweetness standard, sugar (sucrose).
- Other recently developed sweeteners such as neotame (some 8,000 times sweeter than sugar) that has two aspartame components and other bonds, and alitame (2,000 times sweeter than sugar) also have bonds between a few amino acids and aspartic acid as a basic component.
- If we use one of these sweeteners and the bond is broken (by aging, by heat, etc.), a bitter effect will be produced.
- Aspartic acid and phenylalanine can form part of other peptide or protein structures without creating sweetening effects.

Periodic table

scientific concepts

What is it? The classification of all the chemical elements that form matter.

Additional information:

- Recurring properties between the elements are shown, which is how the name periodic table was derived.

	1	2	3	4	5	6	7	8	9
1	1 H Hidrogen								
2	3 Li Lithium	4 Be Beryllium							
3	11 Na Sodium	12 Mg Magnesium							
4	19 K Potassium	20 Ca Calcium	21 Sc Scandium	22 Ti Titanium	23 V Vanadium	24 Cr Chromium	25 Mn Manganese	26 Fe Iron	27 Co Cobalt
5	37 Rb Rubidium	38 Sr Strontium	39 Y Yttrium	40 Zr Zircon	41 Nb Niobium	42 Mo Molybdenum	43 Tc Technetium	44 Ru Ruthenium	45 Rh Rhodium
6	55 Cs Cesium	56 Ba Barium	57 La Lanthanum	72 Hf Hafnium	73 Ta Tantalum	74 W Tungsten	75 Re Rhenium	76 Os Osmium	77 Ir Iridium
7	87 Fr Francium	88 Ra Radium	89 Ac Actinium	104 Rf Rutherfordium	105 Db Dubnium	106 Sg Seaborgium	107 Bh Bohrium	108 Hs Hassium	109 Mt Meitnerium
					58 Ce Cerium	59 Pr Praseodymium	60 Nd Neodymium	61 Pm Prometium	62 Sm Samarium
					90 Th Thorium	91 Pa Protoactinium	92 U Uranium	93 Np Neptunium	94 Pu Plutonium

Elements constituents of food

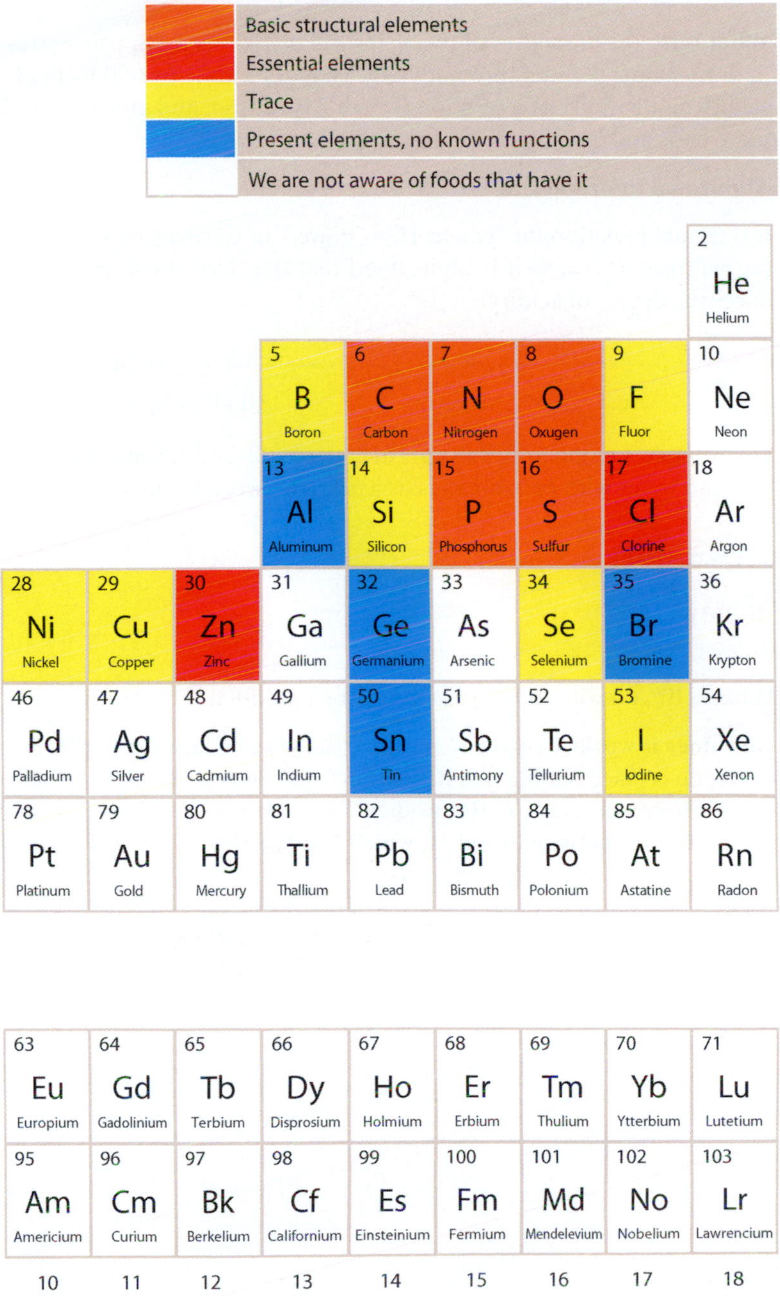
Basic structural elements
Essential elements
Trace
Present elements, no known functions
We are not aware of foods that have it
2 He Helium
5 B Boron
6 C Carbon
7 N Nitrogen
8 O Oxugen
9 F Fluor
10 Ne Neon
13 Al Aluminum
14 Si Silicon
15 P Phosphorus
16 S Sulfur
17 Cl Clorine
18 Ar Argon
28 Ni Nickel
29 Cu Copper
30 Zn Zinc
31 Ga Gallium
32 Ge Germanium
33 As Arsenic
34 Se Selenium
35 Br Bromine
36 Kr Krypton
46 Pd Palladium
47 Ag Silver
48 Cd Cadmium
49 In Indium
50 Sn Tin
51 Sb Antimony
52 Te Tellurium
53 I Iodine
54 Xe Xenon
78 Pt Platinum
79 Au Gold
80 Hg Mercury
81 Ti Thallium
82 Pb Lead
83 Bi Bismuth
84 Po Polonium
85 At Astatine
86 Rn Radon
63 Eu Europium
64 Gd Gadolinium
65 Tb Terbium
66 Dy Disprosium
67 Ho Holmium
68 Er Erbium
69 Tm Thulium
70 Yb Ytterbium
71 Lu Lutetium
95 Am Americium
96 Cm Curium
97 Bk Berkelium
98 Cf Californium
99 Es Einsteinium
100 Fm Fermium
101 Md Mendelevium
102 No Nobelium
103 Lr Lawrencium
10
11
12
13
14
15
16
17
18

pH
scientific concepts

What is it? The logarithm of the reciprocal of hydrogen-ion concentration in gram atoms per liter; provides a measure on a scale from 0 to 14 of the acidity or alkalinity of a solution (where 7 is neutral, and greater than 7 is more basic and less than 7 is more acidic).

Additional information:

It is the abbreviation for "*potenz* H"—"power" in German plus the symbol for hydrogen (because it is understood that the "free" hydrogen ion produces the degree of acidity).

0-------------------------- 7--------------------------14

Acid Neutral Alkali or base

For example: Lemon juice has a pH of between 2.5 and 3, water has a pH of 7, and water with bicarbonate has a pH of between 8 and 9.

pH meter
technology—devices

What is it? A device to measure the pH of a product.

How does it work? In order to measure the pH of a liquid, the pH meter's electrode is placed in the liquid. The electrode detects the presence of ions that indicate the acidity of the product. The pH meter converts this acidity into a number, between 0 and 14, which is called the pH. *See* pH, Acidity.

General uses:

- **In the retail food industry:** Analysis of products and quality control. Control of wine acidity and other products.
- **In restaurants:** In experimentation.

Phosphates

additives—acidity regulators
additives—stabilizers
additives—preservatives
additives—thickening agents
food composition—minerals

What are they? Inorganic compounds with phosphorous that are used as regulators of acidity, stabilizers, preservatives, sequestrants and thickening agents.

Where do they come from? How are they obtained? By transformation reactions of mineral salts.

Form: Powder and the more common granular.

Additional information:

- Acidity regulators: Orthophosphates and disodium dihydrogen phosphate (E-338 to E-443).
- Gasifying agents: By transformation reactions of mineral salts. Phosphate rock is a naturally occurring substance that is mined and purified to form food-grade phosphoric acid. The phosphoric acid is reacted with soda ash to make sodium phosphates, with potassium hydroxide to produce potassium phosphates or with lime to manufacture calcium phosphates.
- E-341 (monocalcium and dicalcium phosphates) and E-450 (disodium dihydrogen phosphate or sodium acid pyrophosphate).
- Thickening agents: Starch phosphates E-1410 to E-1414 (also gelling agents and stabilizers).
- E-450, E-451 and E-452 are mineral additives widely used as pH regulators, as metal sequestrants and as suppliers of electric charge.
- They are very good stabilizers of proteins, ensuring their hydration and thereby preventing them from denaturing at high temperatures.

General uses:

- **In the retail food industry:** Processed meat, poultry and seafood products; dairy products (process or melted cheese); batters and breadings for deep-fried preparations; baking powders; processed potato strips; canned fruits and vegetables and beverages (infant formulas, isotonic beverages and enteral and parenteral feedings). Tricalcium phosphate (E-341) is a useful anticaking agent in powdered mixes (as in spice blends).
- **In restaurants:** Found in food products, no direct use.

Phospholipids
food composition—lipids

What are they? A type of lipids characterized by the presence of phosphorous in their molecules. They form part of the membrane of cells.

Additional information:

They have emulsifying properties in general. For example: lecithin, phosphatydilserine.

Phosphorous
food composition—minerals

What is it? A chemical element that is a component of certain mineral salts called phosphates. Some of its compounds are used as food complements. *See* Phosphates.

Additional information:

- Indispensable to living organisms.
- There are 10 g per kg in the human organism.
- It is present in the energy-accumulating molecules of the organism (known as ATP and ADP). Moreover, it forms part of tissues, bones, etc.
- Phosphorous components are consumed only through food. The most common foodstuffs that contain phosphorous are dairy products (Emmental cheese: 0.6%), egg yolk (0.5%) and pulses (lentils 0.4%).

Physical process
scientific concepts

What is it? A transformation during which the compositions of the substances that make up the product do not change.

Additional information:

- It is considered that these transformations give rise to what may be termed "non-cooked" products.
- The following are examples of physical treatments: cutting, mincing, peeling, grinding, freezing, refrigerating, processing in a blender, etc.

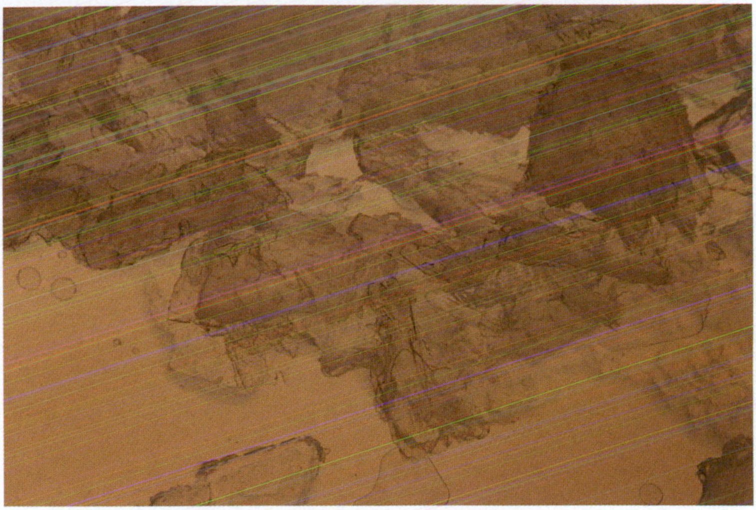

Physics
scientific concepts

What is it? A science that studies phenomena and natural bodies when there are no changes in the composition of the substances of which they are composed, and that investigates the laws by which they are ruled.

Pigments
scientific concepts

What are they? Products that give color to the component they belong to.

Additional information:

- They have strong coloring power and some offer good color stability.
- Many can be used as coloring agents; for example, melanin.

Pipette
technology—devices

What is it? A utensil, generally made of glass, used to precisely measure and transfer small volumes of liquids.

Additional information:

The most frequently used have capacities of 1, 2, 5, 10, and 25 mL.

Poise
scientific concepts

What is it? A unit of viscosity.

Additional information:

- The derived unit, the centipoise (cP), is normally used; 1 P equals 100 cP.
- Recently, the International System has been adopted; 1 cP equals 1 millipascal per second (mPa·s).
- The power of thickening agents is indicated by their viscosity. For example, carob gum at 1% (1 g of carob in 100 g of product) has a viscosity of 3,000 cP or mPa·s, and arabic gum, also at 1%, has a viscosity of 5 cP or mPa·s.

Polymer
scientific concepts

What is it? A large molecule formed by the repetitive linking of hundreds or thousands of smaller molecules called monomers. It is an important macromolecule group.

Examples:

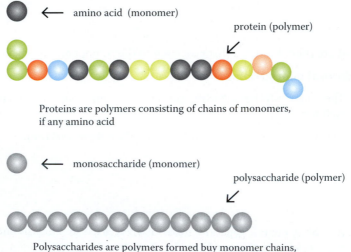

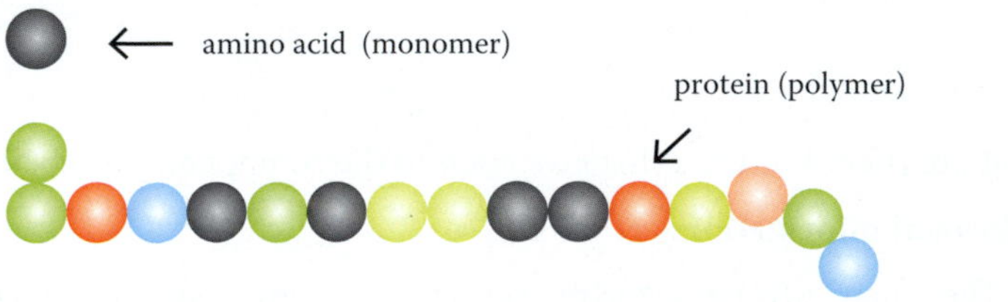
amino acid (monomer)
protein (polymer)

Proteins are polymers consisting of chains of monomers, if any amino acid

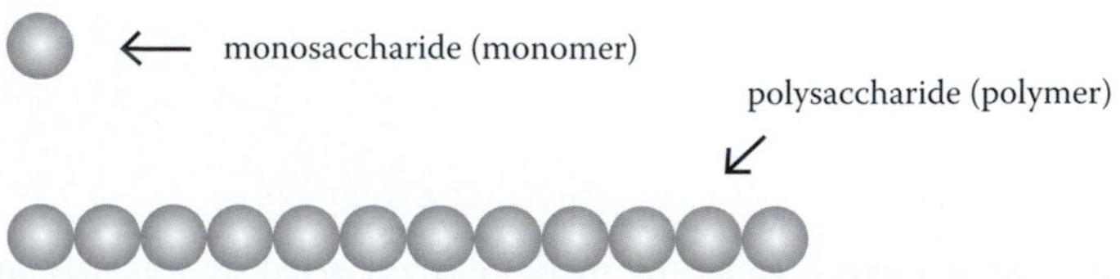
monosaccharide (monomer)
polysaccharide (polymer)

Polysaccharides are polymers formed buy monomer chains, in this case by monosaccharides

Polyols
additives—sweeteners

What are they? A group or family of additive products with a sweetening effect belonging to alcohols.

Additional information:

They have many uses:

- Sugar (sucrose) substitution because they provide fewer calories and do not cause tooth decay.
- Some have the properties of humectants: they delay water loss through evaporation.
- Cryoprotectant: They act as antifreezing agents, allowing the temperature of foodstuffs to be lowered below 0°C without ice forming. Their refreshing effect is also apparent in chewing gum.
- The common disadvantage of all polyols is that they may have a laxative effect in concentrations higher than 60 g/kg.

Types of Polyols:	
Polyol	Sweetening power
Erythritol	0.7 times sucrose
Sorbitol (E-420)	0.6 times sucrose
Mannitol (E-421)	0.6 times sucrose
Isomalt and isomaltitol (E-953)	0.5 times sucrose
Maltitol (E-965)	0.8 times sucrose
Lactitol (E-966)	0.3 times sucrose
Xylitol (E-967)	Equal to sucrose

Polyphenols (food)
scientific concepts

What are they? A group or family of products whose basic chemical components are aromatic groups that are similar to benzenes and that give certain organoleptic properties and colors to various vegetable products, but predominately fruits.

Additional information:

- Flavonoids are within this group.
- The majority of components act as natural antioxidants, which is why they are considered good for health.
- Examples of foodstuffs that contain polyphenols: apples, pears, strawberries, grapes, oranges, bananas, passion fruit, etc.

Polyphosphates (E-452)

additives—sequestrants, chelators

What are they? Long-chain phosphates (sodium and potassium) that are very useful sequestrants or chelators of calcium and magnesium, which gives them the characteristics of stabilizers.

Where do they come from? How are they obtained? From phosphate salts heated to 600 to 800°C.

Form: Powder, granular, flake.

Additional information:

Each molecule is a chain (an average chain ranges from $n = 5$ to 30) of simple phosphate groups. Calcium and magnesium are the common minerals that harden water and cause problems in food processing; sodium polyphosphate is a very useful sequestrant of calcium and magnesium and when used in blends of phosphates optimizes the functionality of other phosphate constituents with the blend.

General uses:

- **In the retail food industry:** Sodium hexametaphosphate (STPP, E-452) is used as a calcium sequestrant in the dairy industry to control alginate gelling, and as a stabilizer in substances called melting salts. When added to cheeses, they enable uniform melting, for example, in fondues. STPP is very useful in maintaining the desired properties (inhibits clouding) of non-carbonated beverages during shelf-life.
- **In restaurants:** No direct use, but found in many foods used in restaurants.

Polysaccharides

food composition—carbohydrates

What are they? Carbohydrates formed by many monosaccharides.

Additional information:

- They can be represented thus:

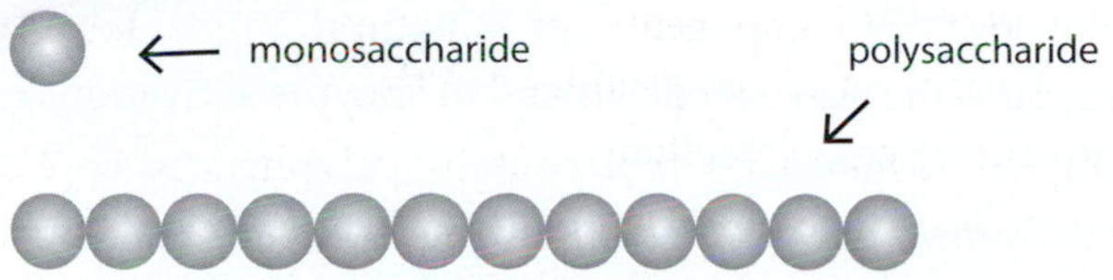
monosaccharide
polysaccharide

- The simple sugars (monosaccharides) that form the chain can all be the same, such as cellulose (formed by glucoses), or they can be different, such as carob gum (formed by mannoses and galactoses).
- There is a great diversity of complex sugars (polysaccharides), and their large dimensions mean that they have very different properties from other smaller sugars (monosaccharides and dizaccharides):
    - They are difficult to dissolve in water.
    - They have hardly any sweetening power.
    - They have no preserving power and are not taste enhancers.
- However, they have the typical properties of hydrocolloids:
    - They are thickening and gelling agents.
- The most important polysaccharides come from vegetable origin (starch, alginates, pectin, etc.). Only glycogen is of animal origin.
- They are classified into digestible (for example, starch) and fiber (for example, cellulose).

See Starch, Fibers.

Potassic salt

See **Potassium chloride**

Potassium

food composition—minerals

What is it? A metallic chemical element that is always associated with other elements (potassium chloride, potassium sulphate, etc.) in the form of salts. It is a food complement and allows the jelling process to take place.

Where does it come from? Mineral salts.

Additional information:

- It is indispensable to living organisms. There are 2 g per kg in the human organism.
- It regulates the osmosis of cells and in enzyme activations related to the breathing chain reactions.
- It is consumed only through food. The foodstuffs that contain most potassium are: vegetables (lentils 0.8%), fruits (apricots 0.3%) and cereals (wheat 0.5%).

General uses:

- **In the retail food industry:** In the form of potassium salts, it has a sequestrant character, is an acidity regulator and produces gelling effects, etc.
- **In restaurants:** No known use.

Potassium chloride

mineral products

What is it? A mineral product known scientifically as potassium chloride that is associated with sodium chloride in some salts.

Where does it come from? From salt mines and the sea.

Form: Crystallized, powder, or granulated.

Additional information:

It has a salty taste that differs slightly from sodium chloride.

General uses:

- **In the retail food industry:** To modify the texture of carrageenans.
- **In restaurants:** Found in some leavening products.

Precipitation

physical or chemical processes

What is it? A gravitational action by which a solid falls through a liquid until it reaches the bottom.

Additional information:

This action can basically take place in two ways:

- **Slowly:** By leaving it for a long time (normally).
- **Quickly:** By centrifuge (increase of gravity by rotation). *See* Centrifuge.

Examples: Fruit juice pulp sinks to the bottom of the liquid. Milk curds that are formed by acid or enzyme actions.

Preservation (procedures of)

food concepts

What are they? Procedures aimed at lengthening the useful life of a foodstuff or to assure its durability for consumption. Procedures may be physical (sterilization, freezing, etc.) or chemical (addition of preservatives, etc.).

Additional information:

- All foodstuffs are altered at varying speeds, depending on composition and environmental conditions (temperature, humidity, oxygen, etc.). These alterations may affect the organoleptic characteristics (color, aroma, taste, etc.) without being harmful to health, or else, in the case of a microbial contamination, they can lead to serious intoxication.
- Some procedures to combat undesired microbes date from ancient times (smoking, salting, curing, pickling, confit in sugar or fat, etc.). The retail food industry uses high temperatures (pasteurization, sterilization, UHT) and low temperatures (refrigeration, freezing), control of free water ($\mathbf{a}_w$), osmotic pressure and other procedures.
- Even so, if there is no assurance that the food will not be contaminated, preservatives may be added (sulphites, nitrates, sorbates, etc.). When used in the permitted dosages, preservatives are not capable of stopping contamination processes once they have begun; they are simply preventive.

Preservation Procedures	
Application of cold	Refrigeration, freezing
Application of heat	Pasteurization, sterilization, UHT
Water extraction	Dehydration, lyophilization
Action of other products	• With salt (salting, brine, etc.) • With sugar (preserves, jams, syrups, etc.) • Smoking, with oils or fats (confits, etc.) • With acids (vinaigrettes, pickled products, etc.) • Action of preservative additives • Others: marinades, selective fermentations
Synthetic additives	Butylated hydroxytoluene (BHT), butylated hydroxyanisole (BHA), tertiary butylhydroquinone (TBHQ), rosemary oil, mixed tocopherols (vitamin E), etc.
Others	Radiation, vacuum packing, packing in inert atmospheres, packing in protective containers, etc.

Preservatives

See **Preservation (procedures of)**

Pressure

scientific concepts

What is it? The amount of force applied to a surface. This concept has been used in cooking processes that are usually carried out when cooking in water inside closed containers.

Additional information:

The different pressure-cooking methods may require an increase in pressure (high-pressure process) or a reduction in pressure (low-pressure process).

High-pressure process:

- An airtight container with a security valve (pressure cooker) is used.
- When water is heated steam cannot escape from the container easily and a pressure that is superior to the normal atmospheric pressure is created.

- This overpressure causes the water that boils at 100°C at atmospheric pressure to boil at between 115 and 130°C, depending on the container. Pressure cookers are designed with a security device that partially eliminates the interior pressure to avoid serious domestic accidents.
- The increase in the cooking temperature by 30°C accelerates by approximately three times the reactions that are produced. The cooking time is therefore notably shorter than it would usually be.
- Obviously, there is water reflow as the steam condenses in the lid and then returns to the boiling mass.
- This cooking method should not be considered merely a variation on thermal cooking with water because reactions that take place at temperatures of up to 100°C are different than those at 130°C. For example, Maillard reactions are initiated at 130°C and carotenes are degraded in carrots, which lose their color as a result, etc. As a consequence, it should be considered as another cooking method with different textures and tastes.

Low-pressure process:

- This is carried out in a closed container that is connected to a vacuum device.
- A vacuum is created inside the container, which considerably reduces the boiling temperature of water. The water could even boil at room temperature, that is, without being heated.
- Experiments can be carried out with this cooking method in places where the atmospheric pressure is low. As altitude increases, atmospheric pressure is reduced. In high-altitude cities (Bogotá, Quito, Mexico City, etc.) the boiling temperature of water in an open container is around 90°C but cooking times are obviously longer.
- The fact that pressure can be reduced with scientific and controlled methods opens new horizons for the retail food industry and gastronomy.

Proteins

food composition—proteins

What are they? Biochemical compounds that contain nitrogen in their molecules and that contribute to the structure of the organism as well as providing nutrients.

Additional information:

- Proteins are groupings of different amino acids that form gigantic structures.
- Their characteristics depend on both the amino acids that they are composed of and their structure. For example, the following are estimates of the percentages of amino acids that are present in collagens and structured in very long chains:

• Glycine (GLY)	35%
• Alanine (ALA)	11%
• Proline (PRO) 1	2%
• Hydroxyproline	(HYP) 9%
• Others	33%

See Amino acids.

- Enzymes are proteins that are responsible for many gastronomic processes that occur at a speed that maintains their viability.
- Proteins are produced only by living matter and their construction is regulated by the genetics of each species according to their basic constitutents, amino acids, as if they were items to construct a perfectly defined building from a genetic point of view.
- Proteins constitute:
  - Most animal tissues together with water; for example, the actine in human muscles.
  - In plants, most proteins are found in seeds; for example, soy.
- Although proteins are often linked to animals, there are examples in the vegetable kingdom with very high proportions of proteins, such as nuts and pulses.

Properties of proteins used in restaurants:

- They are hydrocolloids. This means that they can have different textures with water: they give viscosity, they can gel, etc. For example, collagen can be converted into gelatine.
- Some are emulsifiers (tensioactive). This means that they have the capacity to mix two or more nonmiscible components; for example, ovalbumin.
- They are normally considered to be the "center of the plate" in entrees.

Q

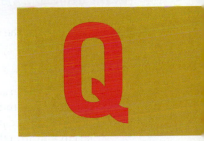

QS (*quantum satis*)
food concepts

What is it? A Latin abbreviation, used in food additives, which indicates that the quantity of a product to be used is the minimum needed to obtain the desired effect.

Quercetin
food composition—pigments and other compounds

What is it? A plant-derived flavonoid present in onions, apples, etc. used as a nutritional supplement. Laboratory studies show it may have antiinflammatory and antioxidant properties.

Quillaia
See **Foaming agents**

Quinine
food composition—alkaloids

What is it? A bitter alkaloid extracted from the bark of plants such as the cinchona tree.

Additional information:

- It is the bitter taste that has been adopted as the bitter standard.

- It is used in the retail food industry for tonics and other non-alcoholic drinks.
- It has been used to treat malaria for centuries.

R

Radiation
scientific concepts

What is it? Electromagnetic energy that is transmitted between two bodies.

Additional information:

- There are many different types: luminous, infrared, ultraviolet, radio waves, microwaves, etc.
- For example, radiation in the microwave that enables substances to be heated.
- The term radiation tends to be used to refer to radioactivity, which is harmful to health in certain doses.

Raffinose
food composition—carbohydrates

What is it? A carbohydrate formed by three linked monosaccharides (glucose, fructose, and galactose).

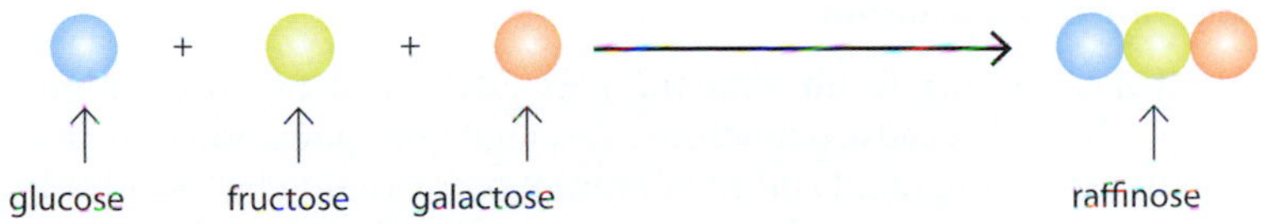
+
+
glucose
fructose
galactose
raffinose

Additional information:

- It is present in plants, especially the sugar beet.
- Raffinose is also found in peas and beans. As it is a sugar that is more complex than the disaccharides, it does not break down easily and enters the small intestine without having been absorbed. It is assimilated (fermented) by the intestinal flora in the large intestine, and these microorganisms liberate gases that can cause flatulence.

Rancidity

organoleptic perceptions

What is it? An organoleptic nuance caused by the chemical alteration of fats.

Additional information:

- The triglyceride molecule is formed by glycerin, which is linked to three fatty acids. When triglycerides decompose, the fatty acid is freed and remains in the oil or the fat, creating a rancid taste. This phenomenon is nearly always caused by enzymes through a hydrolysis reaction (when oxygen initiates, it is called oxidation).
- Wine that has been aged in casks and has acquired undesirable smells and flavors is also called rancid.

Rancidity development

physical or chemical processes

What is it? A phenomenon that often occurs when enzymes cause fat molecules to decompose through hydrolysis. The fatty acids that are oxidized are liberated and produce an off flavor, causing the perception of rancidity.

Additional information:

- It has nothing to do with the peroxidation of fats (formation of products that cause oxidations) although both phenomena are often linked. Antioxidant additives prevent peroxidation but no additives can prevent the development of rancidity; for example, rancidity development in butter, walnuts, ham, etc.
- Unsaturated fatty acids (olive oil, etc.) are oxidized more easily than saturated fats and therefore are more prone to rancidity. Products (aldehydes and ketones) that are formed in rancidity reactions cause the unpleasant smell and flavor.

Reaction

physical or chemical processes

What is it? A process of chemical transformation, in which components (called products) are obtained from other components that are chemically different (called reactives).

Additional information: Most cooking processes are complex chemical reactions between the components of the foodstuffs; for example, the Maillard reaction. *See* Maillard reaction.

Reduction

physical or chemical processes

What is it?

- **Scientifically:** It is the opposite of the oxidation process (in which electrons are gained). Electrons are lost when a molecule is oxidized and won when it is reduced. For example, the reduction of the amino acids in the Maillard reaction gives a characteristic flavor.
- **Gastronomically:** Nothing to do with its scientific meaning. Reduction is when water and other volatile components are evaporated from a preparation or a product in order to obtain a concentrated mixture; for example, the reduction of sauces.

Reflow

physical or chemical processes

What is it? The procedure by which a solution is heated (normally water + one or more foodstuffs) in a closed container with refrigeration that will allow for the evaporated water to recirculate into the solution.

Additional information:

- This method enables all the components of the mixture to be in contact constantly and at a relatively high temperature. For example, wine with sugars on reflow; in this case the alcohol also recirculates.
- Reflow can occur at low temperatures if the container is connected to a vacuum pump.

Refractometer

technology—devices

What is it? A device that offers information on the composition and concentration of products, by means of the measurement of the refraction index, an optical property of matter.

How does it work? The product in dissolution is placed in the device and when light falls on it, the ray of light is diverted (refraction). Information is obtained in accordance with this diversion.

Additional information:

- There are specific refractometers for the following measurements:
  - Sugars (Brix values).
  - Salinity (%).
  - Alcohol content (% volume).
  - Proteins (g/mL).

General uses:

- **In the retail food industry:** Analysis and quality control.
- **In restaurants:** Measurement of sugars (Brix values).

Refrigeration

physical or chemical processes

What is it? A physical process that involves lowering the temperature and maintaining it below 0–4°C. Under 0°C it is a freezing process.

Additional information:

It is used fundamentally for the preservation of foodstuffs during relatively short periods of time; for example, the refrigeration of meat, fish, etc.

Relative humidity (RH)

scientific concepts

What is it? It is the degree of atmospheric humidity, indicated as a percentage of maximum humidity (100%) at a certain temperature.

Additional information: It is measured by devices called hygrometers. It is important to know the relative humidity at which chefs work in kitchens, because this could affect the final product, for example, when working with crisps, caramel, etc.

Rennet

See **Rennin**

Rennin (chymosin)

food composition—proteins

What is it? An enzyme that enables the curdling process in milk to obtain cheeses.

Resin (natural)

scientific concepts

What is it? A solid or pasty substance that is insoluble in water but soluble in alcohol, which is obtained directly from turpentine, oleoresins, or balsams that ooze from certain plants.

Additional information: Some resins, such as oleoresins, have been used in the retail food industry for the manufacture of sweets and as commercial flavorings.

Retrogradation

physical or chemical processes

What is it? A phenomenon whereby the structure of a starch or flor (the film of yeast on the surface of some wines) is modified, releasing water.

Additional information:

- When a starch or flour is cooked in water, a dough is formed that initially has a smooth, shiny surface. After a few hours, this surface begins to crack and form a skin and drops of free water appear. This is due to the amylose component of starch that is well hydrated during the cooking process—once left to rest it begins to establish links and move, without obstacles (because of its smooth structure) toward other neighboring amyloses and to expel part of the water (syneresis). This does not happen with the other component of starch, amylopectin, because the components are prevented from moving due to the amylopectin's branched structure.

- The manufacture of modified starch solves diverse problems created by native (natural) starches, including retrogradation.

Rheology

scientific concepts

What is it? A branch of physics that studies the deformation and flow of matter when a force is applied.

Additional information:

The rheological concepts, viscosity, elasticity, and plasticity are studied when mayonnaise is prepared. The elasticity of chewing gum, spreading ease of butters and margarines, and the fluidity of sauces are rheological characteristics.

Rotavapor

technology—devices

What is it? A device that enables distilled water or alcohol to be obtained (when applied in the retail food industry). Examples include distilled cocoa, coffee, etc.

How does it work? It distills the mixture with rotary movements in a temperature-regulated bath. It can also be connected to a vacuum for low-temperature distillations.

Additional information:

When the device is used at a low pressure the product that remains in the distiller is a reduction that has basically not been cooked at all.

General uses:

- **In the retail food industry:** Recovery of other dissolvents, essential oils, etc.
- **In restaurants:** In experimentation. *See* Distillation.

S

Saccharin (E-954)

additives—sweeteners

What is it? An artificial product derived from benzene that is used as a sweetening additive.

Where does it come from? How is it obtained? In the chemical industry, by synthesis of benzene derivatives.

Form: Powder, pastilles or liquid.

Additional information:

- It has approximately 400 times the sweetness of sugar.
- It leaves a bitter or metallic aftertaste.
- It is one of the most widely used sweeteners in the industry mostly because of its great stability. In fact, most domestic sweeteners are referred to as "saccharin" even though they may be made with other products. Cyclamate and aspartame are widely used at present.

General uses:

- **In the retail food industry:** Confectionery, chewing gum, low-calorie products, products for diabetics, etc.
- **In restaurants:** Products for diabetics.

Saccharomyces

See **Yeast**

Salmonella
scientific concepts

What is it? A type of food poisoning related to the salmonella bacterial microorganism that is sometimes found on poultry eggshells.

Additional information: Some of the ways of avoiding or preventing it are

- Keep eggs cold for use in preparations. Salmonella is not reproduced at refrigerator temperatures (4–6°C).
- Heat the product above 65°C for a minimum of 15 minutes.
- Disinfect the eggshell with acidic or oxidant products: bleach, vinegar, etc.
- Use pasteurized eggs.

Salt
scientific concepts
mineral products

What is it?

- **Scientifically:** A product derived from the reaction of an acid with an alkali.
- **Gastronomically:** The colloquial name for sodium chloride or common salt. It is used as a flavor enhancer, preservative and taste modifier (seasoning).

Where does it come from? How is it obtained? From salt mines or the sea.

Form: Crystallized product.

Additional information:

- There are many types of common salt that are defined by their impurities, added products or the degree of crystallization, but basically they are the same mineral product.
- Some types are
    - Iodized salt, enriched with iodine products.
    - Crystallized sea salt (Maldon sea salt, grey salt, Guérande salt, fleur de sel, etc.).
- Pink salt, which has been formed by the drying up of former sea salt deposits. It contains not only sodium chloride, but also other salts such as calcium, magnesium, iron, and potassium salts that lend it its color.

General uses:

- **In the retail food industry:** As a flavor enhancer and a preservative in meat and fish products, etc.
- **In restaurants:** Used to season most foodstuffs and as a preservative in salted products and brines.

Dosage:

- **Maximum/minimum quantity:** QS (the minimum product quantity needed to obtain the desired effect).
- **Basic quantity for cooking:** QS.

Saponins

food composition—carbohydrates

What are they? A group of slightly bitter products from the vegetable kingdom that are formed mainly by carbohydrate chains that are amphipathic glycosides grouped phenomenologically by the soap-like foaming they produce when shaken in aqueous solutions.

Additional information:

- Licorice glycyrrhizins are noteworthy saponins.
- Saponins are found in low proportions in foodstuffs and as such are not harmful to health, but they are toxic as individual products.

Examples of foodstuffs that contain saponins: Cheese, vegetables (alfalfa, spinach, cabbage, etc.), pulses (peas), licorice, etc.

Saturated solution

physical or chemical processes

What is it? A solution in which the component that acts as a solvent (normally water) does not admit more solute at a certain temperature.

For example:

- 37 g of salt (sodium chloride) are saturated in 100 g of water at 20°C. If more salt is added, it remains in a solid form. More salt can be dissolved in a liquid by raising the temperature until the solution is saturated.
- An example in cooking is brine, which is used to preserve food and is basically salt saturated in water.

Saturation

See **Saturated solution**

Savory

organoleptic perceptions

What is it? The adjective used to describe one of the basic tastes. The product associated with this taste is umami, from meat, mushrooms, and/or cooking salt (sodium chloride).

Additional information:

A savory sensation can also be obtained with potassium chloride, sodium phosphate, and other compounds with characteristics that are similar to sodium chloride.

Sequestrant

scientific concepts

What is it? The name that is given to different types of products capable of capturing another product (totally or partially). The properties of the sequestrated product in the solution are not apparent after the action of sequestrants.

Additional information:

- Sodium tripolyphosphate (STPP) and other phosphates can be used as sequestrants of calcium in dairy products. Although the calcium remains in the solution it does not act as if it were there.

Seroalbumin

food composition—proteins

What is it? A protein of the blood (65% of all proteins, including hemoglobin).

Where does it come from? How is it obtained? From blood.

Form: Powder.

General uses:

- **In the retail food industry:** As a gelling and binding agent.

- **In restaurants:** A component responsible for the gelling of blood sausage.

Silver (E-174)

additives—coloring agents

What is it? An inorganic mineral that is used as a coloring agent for coatings.

Where does it come from? How is it obtained? From physical treatment of silver transformation.

Form: Powder or sheets.

General uses:

- **In the retail food industry:** To coat and decorate confectionery and in baking.
- **In restaurants:** To coat products and preparations, although its use is very limited.

Quantity and instructions for use:

- **Basic quantity for cooking:** QS (minimum quantity required to obtain the desired effect) for the coating of preparations.
- **Instructions for use:** Brush the surface of the preparation with the coating. Mixing with water or alcohol speeds up the process.

Smell

organoleptic perceptions

What is it? The organoleptic sensation produced by volatile particles on coming into contact with the olfactory organ. *See* Volatiles.

Sodium

food composition—minerals

What is it? A metallic element that is always associated with other elements (sodium chloride, sodium sulphate, etc.). Its compounds are used as food complements and to enable gelling processes.

Additional information:

- It is a component of many mineral salts and foodstuffs.
- It is indispensable for living organisms. There are approximately 1.4 g per kg in the human organism.
- It regulates the osmosis of cells and in enzyme activations. Its excessive consumption is related to high arterial pressure (hypertension).
- There are two ways of taking it:
  - Directly, in the form of salt (sodium chloride).
  - By consumption of foodstuffs. The foodstuffs that contain most sodium are cheese, processed meats, and other processed foods.

Sodium alginate (E-401)

additives—gelling agents
additives—thickening agents
additives—stabilizers

What is it? An organic salt derived from fibrous carbohydrates used as gelling and thickening agents and as a stabilizer. It has the properties of a hydrocolloid.

Where does it come from? How is it obtained? It is extracted by physicochemical treatments of brown algae (*Macrocystis, Fucus, Laminaria ascophyllum,* etc.), which are found in cold-water seas and oceans.

Form: Powder.

Additional information:

- For gelling, the sodium alginate needs to react with calcium salts.
- The gel that is formed is thermoirreversible, i.e., on heating it does not return to a liquid state, unlike others such as carrageeenan and fish glue gelatins.

General uses:

- **In the retail food industry:** It has many applications, in particular, the restructuring of cheap and nutritious products that have little attraction for consumers in their original state, giving them a new and pleasing appearance. Examples: derivatives of surimi such as "crab sticks," etc., canned vegetable products (preserves, jellies, jams, etc.), ice creams, etc.

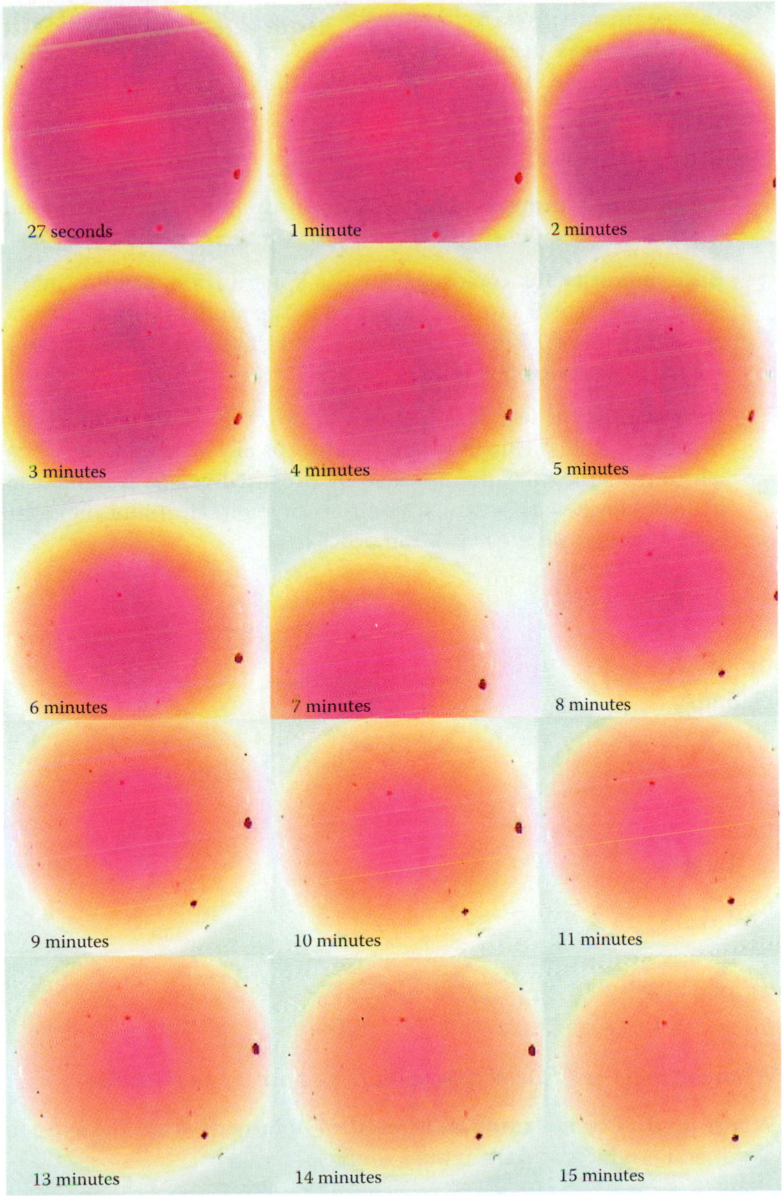
27 seconds
1 minute
2 minutes
3 minutes
4 minutes
5 minutes
6 minutes
7 minutes
8 minutes
9 minutes
10 minutes
11 minutes
13 minutes
14 minutes
15 minutes

Minute-by-minute images of the jelling process of spherical caviar. In the photographs, the pinkish-violet color is the liquid area—the one that has the greatest concentration of sodium alginate. The bright orange is the area in which the calcium chloride is penetrating and producing spherification. As may be seen, after 15 minutes, the sphere has gelled completely. (Photographs taken with a binocular magnifier by Fernando Sapiña and Eduardo Tamayo from the University of Valencia).

- **In restaurants:** As a gelling agent. Its gelling capacity with calcium salts has been used to develop an external gelling culinary technique, invented by elBulli (2003) and referred to as "spherification."

Quantity and instructions for use:

- **Maximum/minimum quantity:** QS (minimum quantity to obtain the desired effect) of 10 g/kg, except in jams, jellies, and preserves.
- **Basic quantity for cooking:** In basic spherication, small proportions of alginate (0.4 to 0.7%) are used in the product with a calcium chloride bath of 0.5 to 1%. In inverse spherification, concentrations are in the experimental phase.
- **Instructions for use:** It is blended by stirring, and heating is not necessary. If it is strongly stirred it absorbs air, which it loses when left to rest. The mixture may also be prepared gently, leaving it to hydrate slowly without causing air absorption. Heat and the presence of sugars aid the hydration process.

Sodium chloride

See **Salt.**

Sodium hypochlorite

See **Bleach.**

Sodium salts

food composition—minerals

What are they? Salts formed by sodium and other components.

Where do they come from? How are they obtained? From the extraction of mineral products and from food sources.

Form: Powder, crystallized, or in water solution.

Additional information:

- The most important example is sodium chloride. *See* Salt.
- Other examples include sodium phosphates, sodium nitrite, sodium alginate, etc. *See* Phosphates, Nitrates and nitrites, Sodium alginate.

Soluble product

scientific concepts

What is it? A product with characteristics that enable it to dissolve in another product that is called a solvent.

Additional information:

Sugar is soluble in water; oils can be soluble in ethanol.

Solute

scientific concepts

What is it? A product that is dissolved by another, normally a liquid (the universal solvent is water).

Solution: Physical or chemical processes

What is it? A homogenous mixture normally composed of two or more (usually liquid) substances that do not react with each other.

Additional information:

- One of the substances acts as a solvent and the other as a solute. The quantity of solute dissolved in a quantity of solvent indicates the composition of the solution and varies with temperature; for example, alcohol and chlorophyll, salt (sodium chloride), and water, etc.
- The differences between a solution, a colloid (emulsion), and a suspension lie in the measurement of particles and stability.

	Solution	Colloid	Suspension
Measurement of particles	Less than 1 nanometer*	Between 1 and 9 nanometers	Greater than 100 nanometers
Stability	Yes	Yes	No

* 1 nanometer is a billionth of a meter.

Solvatation
physical or chemical processes

What is it? A process by which a chemical compound becomes surrounded by other chemical compounds of a product that acts as a solvent.

Additional information:

The process is called hydration when this liquid is water.

Solvent
scientific concepts

What is it? A substance that is capable of dissolving the solute particles in a solution. Generally, the solvent is the majority component of the solution.

Additional information:

- Water is considered the universal solvent.
- Other solvents in the retail food industry include ethyl alcohol, oils, etc.

Sorbitol (E-420)
additives—sweeteners
additives—humectants

What is it? A product from the polyol group that is used as a sweetening and moisturizing additive.

Where does it come from? How is it obtained? By synthesis of glucose (dextrose) and also fructose.

Form: Liquid.

Additional information:

- It is obtained by simple hydrogenation of glucose. It is a raw material used to obtain synthetic vitamin C.
- Sweetening power: 0.6 times that of sugar (sucrose).
- It is found naturally in many ripe fruits (plums in particular, pears, and cherries).

General uses:

- **In the retail food industry:** Chewing gum, coating for candy, and diabetic products.
- **In restaurants:** In confectionery and diabetic products.

Sour
organoleptic perceptions

What is it? The taste that detects acidity. The sourness of substances is rated relative to dilute hydrochloric acid, which has a sourness index of 1. By comparison, tartaric acid has a sourness index of 0.7, citric acid an index of 0.46, and carbonic acid an index of 0.06. The tongue's mechanism for detecting sour taste uses hydrogen ion channels to detect the concentration of hydronium ions that are formed from acids and water. In some publications, this concept is used as a synonym for acid.

Stabilizers
food concepts

What are they? Chemical compounds or mixed compounds that help to maintain the state of a foodstuff.

Additional information:

- They are capable of maintaining a thick, gelled, emulsified texture, color, etc.
- The majority of gelling and thickening agents are also stabilizers; for example, alginates, carrageenans, carob gum, etc.

Starch
food composition—carbohydrates

What is it? A digestible complex carbohydrate (polysaccharide). It belongs to the glycan group and therefore is formed only by glucose chains that may be arranged in linear (amylose) or branched (amylopectin) form. It has the properties of a hydrocolloid.

Where does it come from? How is it obtained? It is extracted from cereals (wheat, corn, etc.) and tubers (potatoes, tapioca, etc.).

Form: Powder.

Additional information:

- Polysaccharides can be broken down and therefore digested due to the presence of amylase enzymes in the organism and glucosidase in saliva and pancreas juices.
- Starch is the most important digestible complex sugar. It is the energy reserve of vegetables. The starch grains swell (hydrate) and break down in water at temperatures of 60–75°C.

- If the product comes from cereals it is referred to only as a starch, but if it comes from tubers it may also be referred to as a fecula. However, it is the same compound.
- The proportions of amylose and amylopectin vary in different foodstuffs and determine their properties.
- Amylose gives the gelled formation prevalence and amylopectin thickens food, increasing its viscosity. For example, if rice is cooked with less than 20% of linear glucose chains (amylose), it becomes sticky, especially if it has been overcooked, as the starch has been broken down (appropriate for making sushi).
- An abundance of branched glucose chains makes the preparation very viscous (it has a thickening effect). For example, tapioca, or exceptionally waxy rice, can be used for thickening with only a small proportion of starch.
- Natural starches have some inconveniences, one being that they form lumps when dissolved. This is why modified starches have been developed. For example, acid treatment produces a modified starch that is widely used in confectionery for the preparation of gelled sweets. Modified starches are additives; for example, E-1404, oxidized starch.

General uses:

- **In the retail food industry:** It has many applications: confectionery, meat products, canned products in sauce, dairy products, pastry, etc.
- **In restaurants:** Cornstarch is the most commonly used thickening agent. However, other starches, such as wheat, tapioca and potato, are important in certain products and cultures. There is also increased use of carrageenans, algaes, etc.

Quantity and instructions for use:

- **Maximum/minimum quantity:** QS (minimum quantity needed to obtain the desired effect).
- **Basic quantity for cooking:** 4% (4 g per 100 g) of liquid to be gelled or to thicken (40 g per kg), depending on the variety of starch.

Instructions for use: Dispersion in cold preparations and hydration in hot preparations, making sure to not overdo it because hydrated starch decomposes.

Types of modified starch:

- **Physically modified starch (with no E number):**
    - Pre-jelled: Already hydrated.

- Thermal: Modified by heat without alteration to their chemical structures.

Chemically modified starch (with E number):

- E-1404, oxidized starch.
- Starch phosphates E-1410, E-1412, E-1413, E-1414.
- Acetylated starch E-1420, E-1422.
- Oxidized and acetylated starch E-1451.
- Other chemically modified starches: E-1440, E-1442, E-1450.

Sterilization

physical or chemical processes

What is it? A process by which most or all microorganisms are destroyed.

Commercial: The process during which most microorganisms, specifically food-borne or infectious pathogens are destroyed to inconsequential levels.

True: The process of complete removal of all bacteria and microorganisms

Processes:

- Conventional sterilization: Heat to 125–130°C for 15–20 minutes.
- Ultra-high temperature (UHT): Heat to 140–150°C for 2–4 seconds.

Still

technology—devices

What is it? A device, usually copper, used to distill alcohol. It can be used to carry out any type of separation by evaporation-refrigeration.

How does it work? After pressing fruit or other products, the juices are left to ferment. Then they are placed in a still, which applies heat to separate the aqueous solution from the high concentration of ethyl alcohol.

Additional information:

- The heated product evaporates and the volatile component, when passed through a (cooling) coil, condenses and is collected in a container. The liquid obtained is very rich in ethyl alcohol and contains fruit essences.
- This device has been used traditionally to obtain liquors.

General uses:

- In the retail food industry: Distilled beverages.
- In restaurants: In-house distillation.

Sublimation

physical or chemical processes

What is it? The direct transformation from solid to gaseous state without transforming to liquid.

For example: Solid carbon dioxide (dry ice) transforms directly to a gas without first transforming to a liquid state.

Additional information:
Sublimation, at low temperature and low pressure, is used to extract water in lyophilization. *See* Lyophilization.

Substitute

food concepts

What is it? It is a product that is used to replace another with similar characteristics. Substitutes tend to be of lesser quality and are prepared when the original product is expensive or cannot be easily obtained.

Additional information:

Substitutes can be

- Synthetic products that substitute for natural products; for example, piperonal as a substitute for vanilla.
- Natural products that are similar to the original products. Examples include chocolate substitute, where vegetable fats substitute cocoa butter; the roe of various fish as a substitute for caviar.

Sucralose (E-955)

additives—sweeteners

What is it? An intensive sweetening additive (discovered in 1986) recently approved by the EU.

Additional information:

- Sweetening power: 650 times that of sugar (sucrose).
- It is derived from sugar (sucrose) using chlorine.

General uses:

- **In the retail food industry:** Baked goods and sweet products. Sugar substitute for diabetics, in biscuits, jams, etc.
- **In restaurants:** Used in sugar-free drinks.

Sucrose

food composition—carbohydrates

What is it? The chemical name for sugar. It is a carbohydrate formed by the bonding of glucose (dextrose) and fructose. The latter provides most of the sweet taste.

Where does it come from? How is it obtained? By physicochemical extraction of sugar cane or beet.

Form: Crystallized or powder.

Additional information:

- It is present in foodstuffs, especially in fruit and vegetables.

Sucrose ester (E-473)

additives—emulsifiers
additives—stabilizers

What is it? An artificial product derived from sucrose (sugar) that is used as an emulsifying and stabilizing additive.

Where does it come from? How is it obtained? From the synthesis of sugar and fatty acids.

Form: Powder.

Additional information:

- Although it is not used much in the United States and Europe because of its high cost, it is frequently used in Japan.

- It has the disadvantages of breaking down at high temperatures and of being expensive compared with other emulsifiers.
- It is digested in the same way as its components (sugar and fatty acids), except for a sucrose ester called "olestra" that contains 6 fatty acids and is eliminated without assimilation; olestra is used as a substitute for fats in low-calorie foodstuffs.
- Its bacteriostatic effect (it prevents bacterial growth) means that it is used to sanitize products; for example, sucrose ester is capable of reaching all parts of harvested vegetables with a large number of creases.
- The most commonly used sucrose esters have a high HLB (between 14 and 16), which means that they are used to prepare oil–water (O/W) emulsions.

General uses

- **In the retail food industry:** It is used in creams, margarines, ice creams (emulsifiers and stabilizers), chocolates (fluidifiers), breads (they increase the duration of springiness), cream (stabilizers), coffee (whiteners), etc. It is also used as a biodegradable detergent.
- **In restaurants:** In experimentation.

Sugar

See **Sucrose**

Sugars

food composition—carbohydrates

What are they?

- **Scientifically:** They are normally associated with simple and double carbohydrates (monosaccharides and disaccharides).
- **Gastronomically:** Products that are used to add a sweet taste.

Additional information:

- Sugars produce a sweet taste to a greater or lesser degree.
- Sometimes sugar is used as a synonym for carbohydrate.

Important Sugars in Gastronomy	
Simple sugars (monosaccharides)	**Glucose** 0.5–0.8 times the sweetness of sugar (sucrose) **Fructose** 1.1–1.7 times the sweetness of sugar (sucrose) **Galactose** 0.3–0.5 times the sweetness of sugar (sucrose)
Double sugars (disaccharides)	**Sugar (sucrose)** 1.0 **Lactose** 0.2–0.6 times the sweetness of sugar (sucrose)
Blends of sugar	**Invert sugar** 1.25 times the sweetness of sugar (sucrose) **Glucose syrup** 0.3–0.5 times the sweetness of sugar (sucrose) **Maltodextrine** 0.1–0.2 times the sweetness of sugar (sucrose)

Sulphites (E-221 to E-228)

additives—preservatives
additives—antioxidants

What are they? Inorganic salts formed by sulphur and oxygen and other elements (sodium, potassium, and calcium) and that are used as preservatives and antioxidants.

Where do they come from? How are they obtained? By reaction between sulphur dioxide and sodium, potassium or calcium hydroxides.

Form: Powder.

Additional information:

- In ancient Roman times, sulphur was burned (giving off sulphur dioxide, which is what performed the task) in wine cellars to preserve wine and cider.
- As gases are complicated to handle and dose, sulphite salts are used instead of sulphur dioxide (formerly known as sulphuric anhydride), as they give off a gas that has antioxidant as well as preservative properties.
- Disadvantages: Some people are sensitive to sulphites (allergies, asthma).

- When digested, they are rapidly transformed into sulphates and are eliminated without causing harm.

General uses: They are very efficient against bacteria and fungi and they have many uses.

- **In the retail food industry:** Dried fruit and vegetables, fruit juices, jams, wines, etc.
- **In restaurants:** No known use.

Surface tension

scientific concepts

What is it? A set of forces that is generated on the surface of a liquid in contact with another medium.

Additional information:

- All liquids tend to form drops because of surface tension.
- This surface tension can prevent an iron needle from sinking in water.
- Emulsifiers (lecithin, sucrose esters, monoglycerides, etc.) reduce the surface tension and allow nonmiscible components to be mixed to form emulsions.

Suspending power

scientific concepts

What is it? A characteristic associated with some products (xanthan gum, iota, gellan gum, etc.) that, when dissolved in water, are capable of suspending solid substances between the surface and the bottom of a liquid.

Suspension

scientific concepts

What is it?

- **Scientifically:** A colloid dispersion of a solid in a liquid (water S/W or oil S/O), or of a solid in another solid (S1/S2).

Gastronomically: A process by which a solid, a liquid, or a gas is suspended in another liquid thanks to the action of a product with suspending power.

Sweetener

food concepts

What is it? Any chemical compound that produces a sweet taste.

Additional information:

Sweeteners can be divided into sugars and sweetening additives. Normally, the name sweetener is reserved only for additives.

Sweetness

organoleptic perceptions

What is it? One of the basic tastes usually associated with sugar (sucrose).

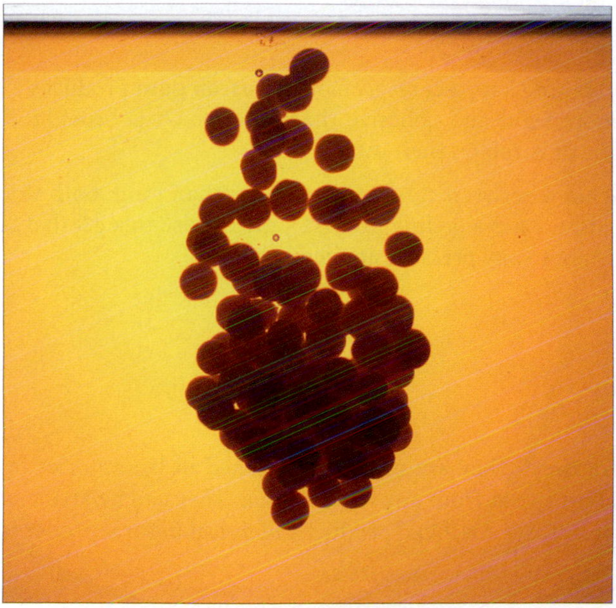

Syneresis
scientific concepts

What is it? The separation or release of a liquid, normally water, once a gelatinous or thick structure has been formed. It is a phenomenon that is related to hydrocolloids.

Examples:

- Some starches absorb water (or are hydrated) only to release some of the water later.
- Another example is products that are gelled with the kappa carrageenan.
- The syneresis phenomenon normally occurs in crème caramel and yogurts, which explains the small amount of liquid that is observed on opening the container.

Synergy
scientific concepts

What is it? An increase in some of the properties of a product by the interaction of one product on another.

Examples:

- Combining kappa carrageenan and carob gum produces a gel that is more rigid and cohesive than that which is brought about by kappa or carob on their own.
- The bond between carob gum and xanthan gum, both thickening agents, creates a gelling property, giving rise to a gel that is highly resistant to any pressure applied.

Synthetic Products
scientific concepts

What are they? They are products obtained in laboratories by means of chemical processes from any raw material. They can be artificial or identical to natural. *See* Artificial products, Identical natural products.

T

Tannins

food composition—pigments and other compounds

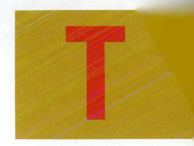

What are they? A set of compounds from the group of polyphenols (associated with flavonoids) that are characterized by giving a certain color to some vegetables and for their astringency.

Additional information:

- They are abundant in nature, present in many vegetable products, especially fruit (grapes, etc.), as well as in cocoa, coffee, and tea.
- Tannins are responsible for the astringent nuance in wine as well as in tea, cocoa, and coffee.
- In fruit, the astringency caused by tannins reduces with ripening.
- Because they are associated with flavonoids, they are considered antioxidants.
- In addition to the retail food industry, tannin extracts are used in tanning to turn animal skins into leather, and in medicine to prepare astringent substances and to treat burns.

General uses:

- **In the retail food industry:** Wines, tonics, etc.
- **In restaurants:** In experimentation.

Tara gum (E-417)

additives—stabilizers
additives—thickening agents

What is it? A vegetable gum from the group of the galactomannans that is used as a stabilizing and thickening agent. It has the properties of a hydrocolloid.

Where does it come from? How is it obtained? It is extracted from the tree *Caesalpinia spinosa* (South America).

Form: Powder.

Additional information:

- The tree it comes from is very similar to the Mediterranean carob tree and the properties of this gum are similar to those of carob gum. This is why it is used as a substitute for carob gum, especially in periods when the price of carob rises sharply.
- It belongs to the galactomannan group, along with carob and guar gum.

General uses:

- **In the retail food industry:** Ice creams, soups, meat, and dairy products.
- **In restaurants:** Used as a thickening agent in soups and sauces.

Tartaric acid (E-334)

additives—acidity regulators
food composition—acids

What is it? An organic acid present in some vegetable products (for example, grapes). It is used as an acidity regulator and a sequestrant of ions that prevent certain processes.

Form: Powder.

Additional information:

- Above all, it is present in the skins of green grapes.
- It is one of the components in mineral water called lithia water.
- As a sequestrant it enables the action of antioxidants.

General uses:

- **In the retail food industry:** Enology, doughs and bakery in general, combined with antioxidants.
- **In restaurants:** In baked goods as cream of tartar.

Tartrates (E-335 to E-337)
additives—acidity regulators

What are they? Products derived from tartaric acid that are used as acidity regulators and sequestrants.

Additional information:

- They have a salty flavor.
- They form part of wine dregs when tartaric acid converts into mineral salts.
- The most important tartrate is potassium hydrogen tartrate, also called cream of tartar.

General uses:

- **In the retail food industry:** Wine making, combined with antioxidants, etc.
- **In restaurants:** Cream of tartar in baked goods.

Taste
organoleptic perceptions

What is it? It is a form of direct chemoreception and refers to the ability to detect the flavor of substances. The sense of taste is directly tied to the sense of smell in the brain's perception of flavor. In the West, the four basic taste sensations are sweet, salty, sour, and bitter. In the Eastern hemisphere, umami (savoriness) has also been traditionally identified as a basic taste. Recently, psychophysicists and neuroscientists have suggested other taste categories (such as fat, metallic, and "water"), although the latter is commonly disregarded due to the phenomenon of taste adaptation.

Tasting
food concepts

What is it? The appreciation of the quality of a product or its characteristics through the senses.

Additional information:

- Widely used in the retail food industry, in the wine and spirits sectors as well as in other products (oils, vinegars, cheeses, etc.).

Tensioactive product

scientific concepts

What is it? A product that lowers the surface tension of water or of a solution.

Additional information:

- In the same molecule one part is soluble in water and one part is soluble in fats, which makes it possible to use it as a detergent and in the retail food industry as an emulsifier and humectant.
- For example: Monoglycerides, lecithin, sucrose esters, etc.

Texture

food concepts

What is it? Physical properties (density, viscosity, surface tension, hardness, etc.) of a foodstuff that give it perceptible characteristics for the senses, especially touch.

Some Possible Textures		
watery	foamy	syrupy
airy	fibrous	doughy
velvety	melting	sticky
soft	gaseous	powdery
soggy	gelatinous	brittle
fleshy	crushed	semi-hard
creamy	grainy	solid
crispy	lumpy	smooth
hard	juicy	unctuous
elastic	liquid	viscous
thick	buttery	etc.

Thaumatin (E-957)

additives—sweeteners

What is it? A protein used as a sweetening additive.

Where does it come from? How is it obtained? It is extracted from the fruits of a tropical plant (*Thaumatococcus daniellii*) that is native to equatorial Africa (Congo, Uganda).

Form: Powder.

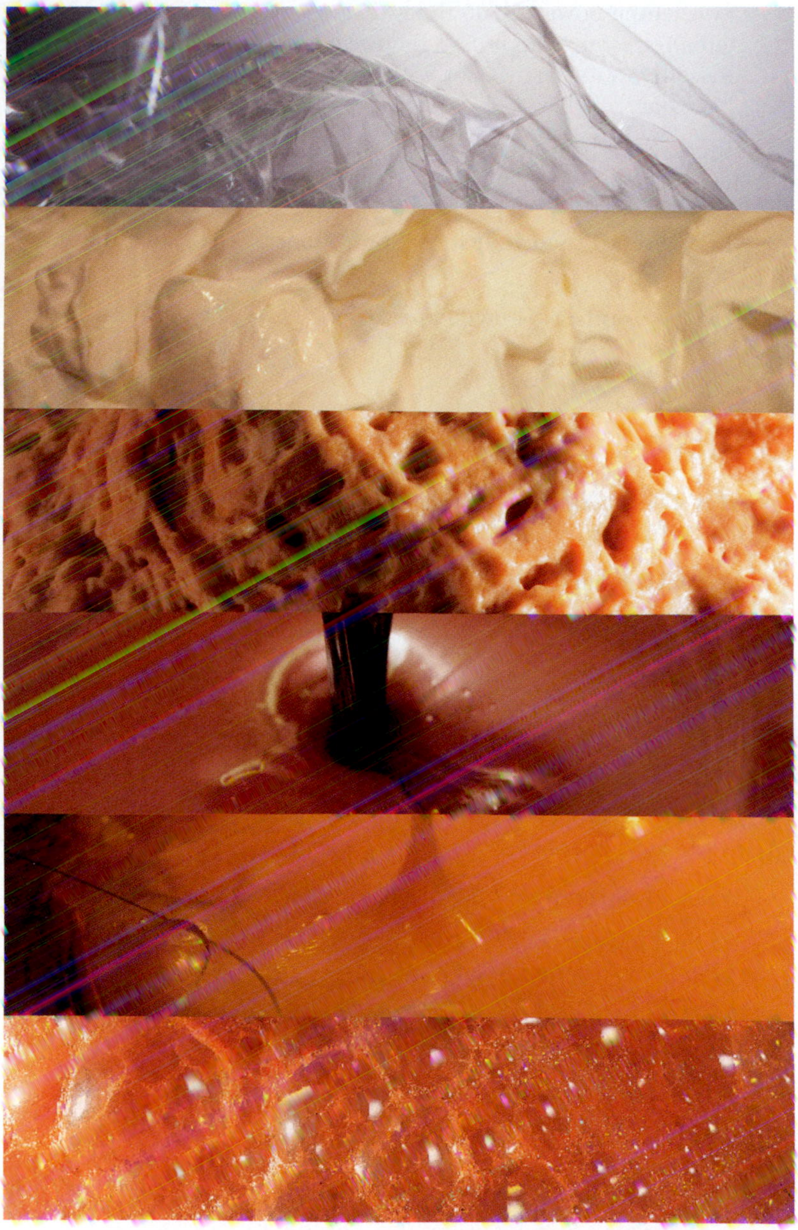

Foodstuffs offer us innumerable textures, whether in their natural state or after handling.

Additional information:

- Sweetening power: 2,500 times that of sugar (sucrose). It is in the *Guinness Book of Records* as the sweetest natural substance known. Recently, tests carried out with neotame gave higher results. *See* Neotame.
- Because it is a protein, it is easily digested.

General uses:

- **In the retail food industry:** Confectionery, chewing gum, dairy products, etc.
- **In restaurants:** No known use.

Theobromine
food composition—alkaloids

What is it? An alkaloid-type product that is present in different plants, in particular in cocoa.

Additional information:

- It can have both stimulating and diuretic effects.

Thermoirreversibility
scientific concepts

What is it? A property whereby once a gel has formed, it cannot be destroyed by temperature.

Additional information:

- In the presence of calcium, alginate produces irreversible gels that keep their structure when heated.
- The HM pectin, which is used in jams and marmalades and for fruit pastes, also provides thermoirreversible gels.
- It is also known as thermostability.

Thermoreversibility
scientific concepts

What is it? A property whereby a gel's consistency is dependent on the temperature.

Additional information:

- Gelatin leaves produce reversible gels. They are jelled at temperatures below 35°C and become liquid at temperatures above 35°C.

Thickening agents

food concepts

What are they? Products that increase the viscosity of a foodstuff in a liquid state. They have the properties of hydrocolloids.

Most Common Thickening Agents		
Fibrous carbohydrates	Algae	Lambda carrageenan
	Microbes	Xanthan
	Plants (seeds)	Tara
		Guar
		Carob
	Plants (resin exuding)	Tragacanth
		Arabic
	Plants (cellulose)	Celluloses
		Carboxymethylcellulose
Starches	Plants (cereals)	Waxy
		Wheat
		Corn
		Rice
	Plants (tubers)	Manioc (tapioca)
		Potato
Flour	Plants (cereals)	Wheat
		Corn
		Rice
		Rye
	Plants (tubers)	Manioc (tapioca)
		Potato

Thixotropy
scientific concepts

What is it? The property of some hydrocolloid gels to reconstruct themselves once destroyed.

Additional information:

When a thixotropic gel has broken down, it can recover its initial homogenous consistency by being left to rest (for example, a gel with the iota carrageenan). This property is of interest to the retail food industry for cold filling (custard, for example) because if the product is filled hot, the water evaporates from the mass and could condense on the lid of the container.

Tocopherols (E-221 to E-309)
additives—antioxidants

What are they? Natural antioxidant additives that are found in the oil of wheat, corn, rice, soy, and even in olive seeds.

Where do they come from? How are they obtained? They are extracted from wheat seeds, rice, etc.

Form: Oily liquid.

General uses:

- **In the retail food industry:** Canned vegetables, cheeses, and fats in general.
- **In restaurants:** Indirectly, when foodstuff is placed in oil solutions.

Tragacanth gum (E-413)
additives—thickening agents
additives—stabilizers

What is it? A fibrous carbohydrate that is used as a thickening additive and a stabilizer. It has the properties of a hydrocolloid.

Where does it come from? How is it obtained? It is exuded from certain bushes (*Astragalus gummifer*) of the leguminous family (found in Syria, Iran, or Turkey).

Form: Powder.

Additional information:

- It was probably used as far back as 2,000 years ago. Thus, it is one of the oldest stabilizers, but little used nowadays.
- It is resistant and has the same thickening property in acidic media.

General uses:

- **In the retail food industry:** Sauces, soups, ice creams, dairy product derivatives, baking.
- **In restaurants:** In frozen desserts.

Transgenic food

scientific concepts

What are they? Genetically modified food.

Additional information:

- They are food that are modified by human intervention to obtain new varieties that result in an increased level of production, plague resistance, profitable components, ease of handling, etc.
- Genetic modification is a normal procedure to improve race by crossbreeding and can be carried out by laboratory manipulation, which is faster and more versatile.
- They are very controversial products. A well-known example is the golden rice variety that contains a precursor to vitamin A. This variety is seen as a prototype for reducing certain malnutrition problems in developing countries.

Transglutaminase (Tgase or TG)

food composition—proteins

What is it? An enzyme that enables the formation of bonds between proteins.

Where does it come from? How is it obtained? From the muscular tissue of fish and mammals. The microbial Tgase can also be obtained from *Streptoverticillium mobaraense*.

Form: Powder.

Additional information:

Transglutaminase creates bonds between the amino acids lysine and glutamine, enabling the link between proteins and therefore the restructuring of certain foodstuffs.

General uses:

- **In the retail food industry:** The reconstruction of pieces of meat and fish to obtain fillets and slices, and in yogurt and cheese production.
- **In restaurants:** Used in binding different meats together and in some dairy applications.

Trehalose

food composition—carbohydrates

What is it? A disaccharide carbohydrate formed by bonding two glucose molecules.

Where does it come from? How is it obtained? By treatment of starch.

Form: Powder.

Additional information:

- It is considered to be a source of glucose.
- It is a type of sugar that has been commercialized in Japan for years and received approval of the American Food and Drug Administration (FDA) and recognition of GRAS (quality certificate) in the year 2000. Its commercialization was approved in Europe by the EU in 2001.
- Among the most important properties of trehalose are its capacity to sweeten approximately half as much as sugar, and the protection that it offers to membranes and proteins in the process of drying or freezing.

General uses:

- **In the retail food industry:** It is used to protect structures, for example, the proteins in surimi. It is also used as a protective barrier against moisture.
- **In restaurants:** In experimentation.

Triglyceride

food composition—lipids

What is it? A molecule formed by glycerin bonded to three fatty acids.

Additional information:

- All oils and fats are blends of triglycerides; the difference between each one lies in the fatty acids that intervene (oleic, palmitic, stearic, linoleic, lauric, etc.).
- The triglyceride will have certain properties, depending on the fatty acids present (solid or liquid, whether it can be easily oxidized or not, taste, etc.). *See* Fatty acids.

Trisaccharides

food composition—carbohydrates

- **What are they?** Carbohydrates formed by three monosaccharides.
- **Examples:** Maltotriose, panose, raffinose, etc.

U

UHT (ultra high temperature)

physical or chemical processes

What is it? A type of sterilization that is used on a product at a temperature of 135–140°C for 1–5 seconds, with the aim of eliminating most of the microorganisms that are present.

Additional information:

Much used in milk, but it can be applied to various products, e.g., juices, creams, wine, soups, and even prepared dishes.

Umami

organoleptic perceptions

What is it? According to Japanese taste classifications, it is one of the basic tastes. It is associated with a mineral (metallic) sensation in the mouth, due to the presence of sodium.

Additional information:

- Basically associated with glutamate (sodium glutamate), although there are other substances to which it can be attributed, i.e., inosinate and guanylate.
- Linked to Asian cooking and an important component in sauces, such as soy sauce. Although the Japanese have acknowledged it for years, it has still not found wide acceptance as a basic taste in western society.

V

Viscometer
technology—devices

What is it? A device to measure the viscosity of a liquid.

Additional information

One of the most commonly used is the Brookfield type.

General uses:

- **In the retail food industry:** Used in experimentation and quality control to find the degree of viscosity in soups, purées, jams, sauces, etc. Also used in hydrocolloid food applications and for checking viscosity.
- **In restaurants:** In experimentation.

Viscosity
scientific concepts

What is it? A material's resistance to flow.

Additional information:

- Normally it is associated with liquids (fluids).
- Viscosity should not be confused with density; for example, oil is less dense and more viscous than water.
- It is normally measured in millipascal-seconds or centipoises (cP), with a device called a viscometer.

Vitamins

food concepts

What are they?

- **Scientifically:** Biochemical products that cannot be synthesized by the body that are vital for the correct functioning of the organism.
- **Gastronomically:** Substances contained in foodstuffs that are indispensable for optimum use of their nutritional principles.

Additional information:

- The organism should obtain them from foodstuffs (vitamin A from carrots, for example).
- Three vitamins are also used as additives: B2 (E-101) as a coloring agent, and C (ascorbic acid, E-300) and E (E-306) as antioxidants.

Volatiles

scientific concepts

What is it? Molecules that evaporate easily at normal or cooking/eating temperatures and can be perceived by the sense of smell.

Additional information:

- Volatile molecules can be odorless (water, carbon dioxide, etc.) or odorous. In the latter case, the effect is produced because the molecules link up to the olfactory receptors that are found in the nose. According to recent studies, we are capable of differentiating some 10,000 different scents.
- When the receptors are linked to the molecules of the volatile compounds, they transmit the information to the brain by means of the nervous system. If this were to be studied and controlled there would be a revolution in the worlds of perfume, aromas, and gastronomy.
- Basically the types of particles (molecules) that cause these sensations are
    - Esters: For example, methyl anthranilate gives smell and taste to grapes.
    - Essential oils: For example, limonene that is present in lemon essence.
    - Aldehydes: Present in smoked fish and meats.
    - Ketones: Offer flavor in cooked meats.

- Acids: Butyric acid gives the scent of butter, and aged cheeses.
- The most common types are present in foodstuffs, often combined to give nuances that determine a certain smell. Sometimes the combination of many products is necessary to create the sensation of smelling a particular food product.

W

Water

food composition—minerals
mineral products

What is it?

Scientifically: The formula H_2O indicates that it is a chemical compound of hydrogen and oxygen. It forms part of nearly all foodstuffs, except oils, salts, and sugars. Examples: a pork chop is 60.6% water, a hen's egg is from 70 to 75%, a pear is 84.7%.

Uses: A consumer product much used in different preparations. In reality, the water that we consume is a very diluted mixture of water (H_2O) and mineral salts. Water bottled directly from springs without being in contact with the air is often called mineral water.

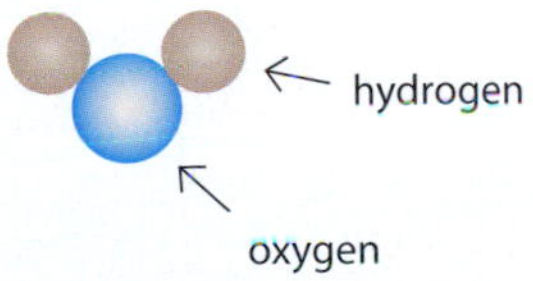
hydrogen
oxygen

Representation of the water molecule.

Percentage of Water in Some Products	
Product	**Percentage**
Milk	85–90
Hen's eggs	70–75
Oils	Practically nil
Fish	70–80
Meats	65–75
Cereals, flours, starches	10–15
Fleshy fruit	70–90
Nuts	5–6
Vegetables	80–95

Water activity (a_w)
scientific concepts

What is it? It is the "available" or "free" water in a food product.

Additional information:

- A part of the water contained in food is strongly "bonded" (linked with or forming part of structures such as crystals, capillaries, proteins, etc.), whereas another part is available or "free" ($_{aw}$). Microbes survive only in free water conditions. The greater the quantity of free water, the greater the possibility of microbe contamination.
- Despite this, it is a necessary factor for the multiple chemical reactions that take place in food; some are positive (maturing of cheese, cured meat) and others negative (oxidation of fats).
  - If the $_{aw}$ of a food is known, the viscosity of its juice may be anticipated before blending. For example, it will provide information as to whether a juice can be prepared from processing fruit in a blender or if water should be added.

Whey
food concepts

What is it? A residue (murky liquid) that is left over when milk is curdled to prepare cheese.

Additional information:

- Nowadays it is used for human nutrition, but for years it was discarded as waste (causing pollution in rivers) or used as a drink for animals.
- Liquid whey is made up of 95% water and 5% lactose and soluble proteins (lactoalbumins, lactoglobulins) that are becoming increasingly recognized because of their capacity to gel as well as for their nutritional value.
- The water is evaporated and the whey is obtained in powder form. This whey is used in desserts, biscuits, etc. The lactose, soluble proteins and minerals, etc., can be separated by ultrafiltration, to considerably increase the application possibilities.
- The whey proteins are heated to 85°C to separate (precipitate) the proteins that did not coagulate with the rennet. Examples of whey cheeses are ricotta, Brunost, Monouri and Urda.

General uses:

- **In the retail food industry:** Dairy product desserts, biscuits, etc.
- **In restaurants:** Powdered whey is used in ice creams, gels, and in smoothies.

White lichen

food concepts

What is it? The name given incorrectly to certain algae when cleaned, blanched, and dehydrated.

For example: *Chondrus crispus (Irish moss)*, from which carrageenans are extracted.

X

Xanthan gum (E-415)

additives—thickening agents
additives—stabilizers

What is it? A fibrous carbohydrate that is used as a thickening additive and a stabilizer. It has the properties of a hydrocolloid.

Where does it come from? How is it obtained? It is produced by the fermentation of cornstarch with a bacterium (*Xanthomonas campestris*) that is present in cabbage.

Form: Powder.

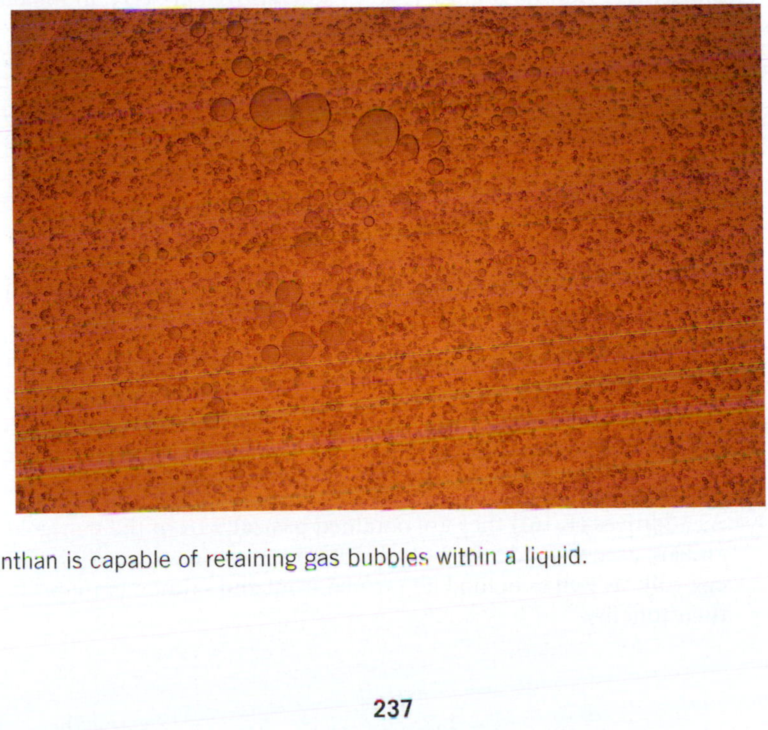

Xanthan is capable of retaining gas bubbles within a liquid.

Additional information:

- It cannot form gels on its own but it can give viscosity to the foodstuffs to which it is added.
- It is stable in a very wide range of acidity. It is soluble when cold and hot and it resists the processes of freezing and thawing very well.
- It gives an elastic gel when mixed with carob gum at a ratio of 1:1.

General uses:

- **In the retail food industry:** Emulsions, such as sauces, ice creams. A stabilizer of beer froth. It is capable of forming gels when mixed with other polysaccharides, especially carob gum, and as such it is used in puddings. It is also used to give consistency to low-calorie products.
- **In restaurants:** In sauces, soups, desserts and foams as well as in experimentation.

Xanthophylls (E-161)

additives—coloring agents
food composition—pigments and other compounds

What are they? Natural orangy-yellow pigments that are very abundant in nature (eggs, flowers, fruit, herbs) and used as coloring additives.

Where do they come from? How are they obtained? They are directly extracted from natural products (alfalfa, tomato, etc.) or by chemical synthesis.

Form: Powder or oily liquid.

Additional information:

- They are responsible for the yellow color in vegetables, although they are often hidden by chlorophyll.
- They give color to egg yolk, as well as some color to salmon flesh and to the shells of crustaceans. In the latter case, the xanthophyll is linked to a protein and acquires a bluish or greenish color; this bond is broken down on heating, which explains the change of color of crustaceans when cooked.
- As additives (E-161) they are obtained basically from the marigold and are used in fodder for hens to obtain a more intense color in the egg yolk, as well as in food for farmed trout and salmon to intensify their tonality.

General uses:

- **In the retail food industry:** Ice creams, margarines, sauces, confectionery, drinks, etc.
- **In restaurants:** Confectionery.

Xylitol (E-967)

additives—sweeteners

What is it? A product from the polyol group that is used as a sweetening additive.

Where does it come from? How is it obtained? By synthesis of cellulose and other vegetable products.

Form: Powder.

Additional information:

- Sweetening power: Equal to sugar (sucrose).
- Although it is manufactured by synthesis, it can be found in nature in fruit, vegetables and cereals.
- It is a widely used product in toothpastes, as its characteristic refreshing effect provokes an increase in the production of saliva that helps clean and protect teeth; moreover, it reduces the growth of microbes associated with tooth decay.

General uses:

- **In the retail food industry:** Chewing gum, chewing gum that acts as a mouth freshener, coating for candy and diabetic products.
- **In restaurants:** Soft pastries and "cooling" desserts.

Y

Yeast

scientific concepts

What is it? The name given to a group of unicellular fungi that is abundant in nature and used to cause fermentations. *See* Fungi.

Additional information:

- The retail food industry is interested in yeasts that are capable of making doughs ferment, of making sugar ferment when producing alcohol and carbonic gas, and that produce the fermentation of beer, etc., and that are capable of preventing spoilage yeasts.
- Saccharomyces are the most important and almost the only type of yeast. Their name comes from the Greek words *sakchar* meaning sugar and *mykes* meaning fungus. They are involved in fermentation processes that allow the production of bread and alcoholic beverages. During the process, carbon dioxide gas is given off, resulting in springy bread doughs and the presence of dissolved gas in some beverages.

See Microorganism (or Microbe).

Theme Index

Additives

Theme Index

Food composition

Food concepts

Mineral products

Organoleptic perceptions

Physical or chemical processes

Scientific concepts

Technology